SALES FOR SCIENTISTS

BY JASON GUSS, PHD
CEO | FOUNDER | RESEARCHER | SCIENTIST
AND
STUART NIXDORFF, MBA
ENTREPRENEUR | DEEP TECH MARKET MAKER
RIGHT HAND, GO TO MARKET MENTOR FOR
TECHNICAL FOUNDERS

Copyright © 2023

All rights reserved. No part of this publication may be reproduced, distributed, or transmitted in any form or by any means, including photocopying, recording, or other electronic or mechanical methods, without the prior written permission of the publisher, except in the case of brief quotations embodied in critical reviews and certain other noncommercial uses permitted by copyright law.

Book Design by HMDpublishing

CONTENTS

Preface 4

Chapter 1
A Primer on Communication 17

Chapter 2
Perspective - the Secret Skill to Selling Success 32

Chapter 3
Credibility - The Technical Hammer 52

Chapter 4
Framing 68

Chapter 5
Creating Momentum & Urgency 85

Chapter 6
Presentations 95

Chapter 7
Storytelling 114

PREFACE

My Worst Fear Unimagined Come True

It was March of my final year of my Ph.D., and I was attending and speaking at the Orthopedic Research Society annual event in New Orleans. After five years of preparation, thousands of hours of research, and an unbelievable amount of planning, I was finally ready to present my research. I sat in the cold lobby of the Hilton, not too far from Bourbon Street, frantically scrolling through my presentation slides. In less than 30 minutes, I would be in front of nearly 500 people... just me, the podium, and my slides. This event didn't only boast the leading researchers in my field, it was also the culmination of my Ph.D.

I had run through the slides countless times, rehearsing over and over, but still, nervous energy was running through me like static electricity. What if I got on stage and I forgot my name? What if I forgot what I was supposed to say? What if I just went silent? What if the crowd laughed me off stage? My mind was going through a seemingly endless loop of "What ifs."

I couldn't wait to get this over with. Just before I arrived where I was to present, my friend Derek passed by and shouted, "Good luck, Jay. You're gonna crush it!"

I wasn't so sure. I grabbed my laptop with one last glance at the title slide. *I got this... I hope so.* I exhaled, closed my bag, and stuffed my computer into my bag.

As the presenter before me on the agenda finished, my feet were tapping on the ground anxiously. Only a few seconds separated me from my fate. I was then announced as the next speaker, and the room fell silent. I weaved my way through rows of chairs with little legroom, and the podium, only several feet away, now felt like an insurmountable summit. Finally, on stage, I shook the moderator's hand. He looked cool and calm, probably because he didn't have to present. I gazed across the room at my peers and my advisor in the front row and then grabbed the PowerPoint remote on the podium. But as I turned to the screen to see my title slide... there was absolutely NOTHING. A big blank screen resembling the black hole I felt in my chest. "Where are my slides!?" I asked the moderator

"The slides will be up there shortly," the moderator said, flashing a nervous smile.

The technician then came up to fix the file and PowerPoint, but for some reason, it wouldn't load.

"Did you upload it?" the moderator yelled across the audience.

Yes, yes, I did; twice and I emailed a backup as well. My skin was crawling by being asked these obvious questions. Did they think I was an unprepared presenter? Did the 500-person audience that included future potential bosses and colleagues think so?

"We are running out of time," the moderator called out from across the room. "We are on a tight schedule."

"I can't get it," the technician shrugged.

The moderator urged me in front of everyone, "You need to start your presentation without slides, or the entire day will be off track."

I looked back at the audience, both of us in shock that I would have to do this without slides. They gulped, and so did I.

Well, folks, this is right about where my head exploded. The months I spent crafting a perfect presentation and the five painstaking years performing research and analysis all felt for naught. This didn't even happen in my worst nightmare of the hundreds of scenarios I played out in my head in advance. I had to get out of my head and give some kind of presentation, and quick, because 500 people and my career were depending on it. It was time to put up or shut up. I took a look at the crowd, back at the blank screen, then back at the crowd. "Good afternoon, everyone. I'm Jason Guss, and the title of my talk is......". And the rest is now history.

That afternoon, I gave an impassioned speech, focusing only on the important details and points as I no longer had images to reference. I inserted several jokes, referencing and pointing to the figures on my slide that were not there. And by the end, the audience gave uproarious applause. So much so that the other presentation halls heard it, and several people I spoke with later said, "that was you!" The next day, I found out that not only did my talk go well, but *I won the award for best speaker.* Would you believe that? I turned lemons into lemonade. But doing so did not happen magically. It was a result of extensive preparation, focus, and a focus on my sales, communication, and verbal abilities. Luckily, I had practiced countless times. I had crafted a story that had a logical flow for both my audience and for my brain to follow without slides. And I had practiced my communication skills frequently and could see through the lens of the audience's eyes.

The point of opening with this story is two-fold:

1. To start with what was hopefully an engaging and relatable story with high stakes. As even the most prepared experts and speakers will get nervous before their talks. It's completely natural, and you can and should use the extra energy to deliver an exciting and engaging story. Remember, that energy has to go somewhere; you might as well use it for good. I always tell myself before a talk that the energy I'm feeling is not nerves but *excitement*. It gives it a nice label and is a healthy way to reframe the discomfort into healthy, tangible action.
2. To show you that if you devote time to selling and communication, you can overcome even the worst possible

handicap. In this example, none of the countless fears I worried about came true; it was something completely different! So don't waste time in your head paying attention to "what if" scenarios because one worse than I could have imagined came true, and I did better than I could have imagined. And you can too!

Hello from Us

We are so happy you decided to make an investment in sales by picking up this book. If the idea of selling makes you a little (or a lot) uncomfortable, that's good; it means you're about to grow.

This book is written by a scientist turned salesperson and a deep tech sales expert with decades of experience. Between the two of us, we've built and sold companies, sold to and worked for Fortune 500 companies, have published award-winning research at Ivy league schools, and worked as and for an engineer and Ph.Ds. The scientist in this book started out shy, with limited to no communication or presentation skills, and most certainly no sales skills. With dedication and more than a decade of effort, he went from being too shy to present to one person in a room, to commanding the stage in front of thousands. The salesperson started with an interest in sales with limited innate skills early on, but had a passion for technology and science. With dedication and a full career of selling, the salesperson learned every trick of the trade, created some new ones, and partnered and mentored many scientists in his career.

Together, we've created a Guidebook, Cheat Sheet, and Accelerator to provide you with the most impactful and critical secrets of selling in as short a time as possible. In this book, you will hear firsthand accounts and learning experiences from both the scientist and salesperson that will prepare you for key sales situations you will face. The book will help provide you with fundamental secrets, hacks, and necessary skills to learn the art of selling, as well as dozens of exercises to practice, so you can succeed and grow, both in your professional and personal endeavors. We want to save you the decade it took us to learn all this; consider this the shortcut to learning sales so you can get what you want and deserve faster.

Who Is This Book For?

This is meant to be a handbook for any aspiring scientist, engineer, science enthusiast, technologist, or person generally looking to pick up a few essential tips on communication and sales. There are many self-help and sales books out there, but NONE seek to teach the skills of sales specifically to people in STEM careers and the sciences.

The book can be used by anyone at any stage of their career, though I wish I had access to this information as early as high school—it would have gone a long way in college or during my Ph.D. Unfortunately, for the majority of STEM careers, there is a gap in the education of sales skills, both in industry and academia, leaving many without these indispensable life lessons. Some may acquire these skills through brute force or via trial and error as a requirement for their role; some may even be so fortunate to possess a natural aptitude toward sales, but most never learn, much to their detriment.

Are you...?

- A currently enrolled or recently graduated Ph.D. student looking to pursue a career in science but don't have any formal training in presenting or communication?
- A researcher at a university, lab, or research center looking to break above the slow growth path from researcher to level 2/3/4 researcher?
- A professor at a research university looking to improve the impact and dissemination of your body of work?
- An engineer at a firm or company freshly out of an undergraduate program looking to prepare for the business world and career growth?
- An entrepreneur looking to build and fund a scientific-related endeavor but with limited experience in pitching, selling, or fundraising?
- A student pursuing science or engineering as an undergraduate and looking to stand out from the other students while complementing your science/math skills?

- A technology manager at a Fortune 500 company looking to have more of an impact in your role and have their voice recognized by leadership?

If so, then this book is for you.

Often, many scientists, when wanting to change roles, jobs, or careers, realize the need for new skills. Learning how to sell will increase and accelerate your ability to move into and with your desired role, job, or career change. Sometimes, professionals may even go for an MBA, but even these expensive and lengthy degrees do not teach the skills and secrets taught in this book. You can choose to get the MBA, or depending on your goal, you learn the high-impact skills that accelerate getting where you want to go right now.

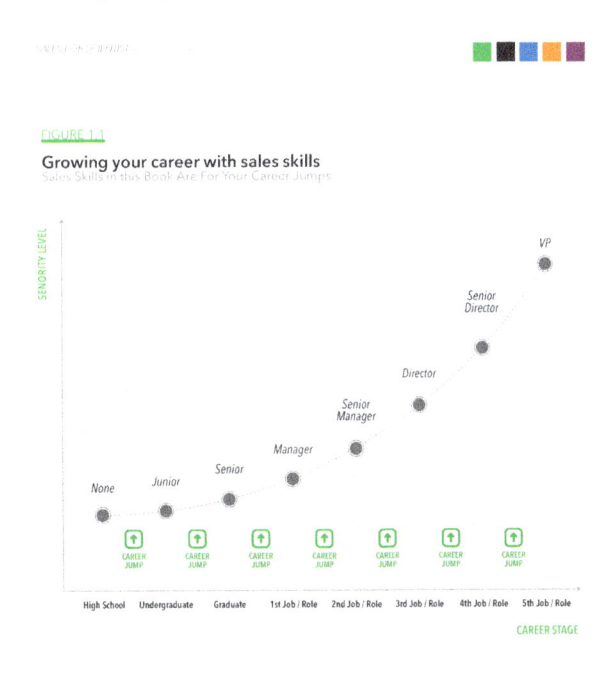

FIGURE 1.1

Growing your career with sales skills
Sales Skills in this Book Are For Your Career Jumps

Figure 1. Sales Skills Help You In Every Step and Inflection Point of Your Career

Everything inside this book will help provide you with tangible skills and support your growth as a communicator, influencer, storyteller, and seller.

Throughout this book, you will see the archetype of who this book is for described as "scientist". This is simply shorthand to describe you, our target reader.

How to Get the Most Out of this Book

Before starting this book, make sure to think long and hard about your goals. What do you want to improve on as a scientist, engineer, communicator, and storyteller? Where do you want to be in your career several years from now? Why are you looking to change and improve? Without a clear goal, reason, and plan to execute, the lessons in the book may go to waste. Be ready to put in the work to improve as you read the book. If you are not willing to put in a little effort right now to improve yourself, then right now may not be the time to read and work through this book.

To get the most out of the book, there are a few simple things that can help from the start:

1. Write down your clear goals for the book. Make sure to list your desired improvement(s) and where you would like to be 1, 5, 10, 15 years from now.
2. Write down your "WHY." This is incredibly important to maintain motivation and the reason to improve.
3. Set Performance Milestones - steps toward your desired outcome.
4. Use and create metrics and timelines to track your progress toward your goals. For example, if you notice a habit you would like to change, start taking a note every time you perform that habit. Start tracking and making changes today. For any habits you want to change, take a note on how often they happen.
5. When you approach exercises in this book, make sure to spend 30 - 60 minutes on them at that moment. It will only help improve your abilities, your branding, your messaging, and your communication skills.

6. As you move through the book, make sure to take notes in a separate document. Write down pointers and examples you can use in your life, any words of wisdom that stick out, and any other lessons that stand out to incorporate into your life.

EXERCISE: WRITE DOWN YOUR GOALS AND YOUR WHY

I know I'm already asking for work, and you just got started… but this is important! For this exercise, I want you to take 10 minutes and write out one clear, overarching goal for this book. It can be "Help land my next job," "To win a presentation award," "To build my own brand," or any goal related to communication or otherwise. But at least write down one goal. Once you have written down your goal, I want you to then write down WHY you want to achieve that goal. Make sure to dig deep and try to understand what is motivating you to achieve that goal. Use the blank space below to fill it in:

GOAL:

WHY:

The Attack on Science: Data and Facts Don't Matter; Stories Do.

The attack on science and scientists is an unfortunate byproduct of the fear of change and the over-politicization of science in today's society. Staring into the abyss of a global pandemic with family members dying and the wake of the end of the planet as we know it, we still find ourselves begging to get the general public to believe, trust, and accept science.

With the recent global Covid-19 pandemic, the spreading of lies, the dethroning of facts ("fact" being the term for an observation repeated enough that it's accepted as "true"), and the blatant disregard for science has never been stronger.

So how do we move past this? How do we preserve the scientific facts, achievements, and legacy that have brought us forward as a country and as a civilization to new frontiers? How do we pair this with the most basic principles of humanity? That we are susceptible to stories and our emotions, and this is what drives and motivates us at our core.

To get the world to accept and believe in science once more, we need our front-line scientists, professors, and researchers to improve themselves as storytellers. The good news is this can be taught and improved upon with practice and education. Storytelling strategies, methods, and practices are discussed in depth later in this book. If we don't learn to tell stories about who we are and what we do, and if we don't learn to tell them well, then we risk becoming a modern-day Galileo or Copernicus, our work and the fate of humankind sentenced to metaphorical or real death.

The Status Quo in Science

FIGURE 2. A STEREOTYPICAL SCIENTIST

The lead actor of the (fictional) film *The 20th Century Scientist* walks in; he (of course, it's a "he") becomes nervous and can barely speak. He is super awkward, to say the least. Our unlikely leading man is extremely smart and is a stereotypical "nerd" that has trouble communicating ideas.

This is pretty much the stereotype of a scientist we have seen or heard for decades on TV, movies, radio, and anywhere else. I remember watching James Bond or other action movies growing up, and whenever a scientist appeared, you could almost guarantee he would be portrayed as a nerdy, awkward stereotype (a stereotype that still exists today). And as a kid who grew up to be a scientist and receive a Ph.D., I also had an expectation of this stereotype.

The current system producing engineers and scientists also supports this, with limited formal training to be better communicators and storytellers. Scientists come in all shapes and sizes, all races and ethnicities, genders, and religions. Scientists are amazing, funny people, who are friends, family, loved ones, and more. We need to start showing this to the world. And we need to start teaching scientists how to move away from the stereotype so they can thrive as the unbelievable communicators and innovators they are.

Case in Point

When we were selling our technology at my startup into the food processing industry, we learned that there was an undercover bias and impact of having a team with high-level degrees speaking to those who did not have the same level of education as us. Meat processing is an industry historically less technology forward, and where many, if not the majority of workers, do not have high levels of education beyond high school. We were presenting the insights and findings from our analysis to a management team, and so we made our presentation as detailed, jargony, and scientific as possible. After all, we wanted to sell that we are expert data scientists and only we could look at these complex analyses. As the project progressed, we would present our findings at our weekly stakeholder meeting. At every weekly presentation, though, we were not getting any feedback or questions on what we thought to be impressive results. We were so worried to the point that we thought they may not be interested or find value in our offering. Finally, we were able to get one of the members in the group to speak with us "off the record." Surprisingly, what we found out was that the managers did not feel confident to comment or ask questions on anything because we are all Ph.Ds with advanced degrees, and the fear of sounding stupid was real. Rather than impressing them with our amazing findings and degrees, instead, we almost lost the deal. This story and experience has stuck with me and made me aware that the Ph.D. and fancy degrees can be seen as a scarlet letter, so to speak. Rather than impressing those with our degrees, we may scare them away, and make them deaf to our data. So, in this case, we saw this as our chance to make them laugh and bond. Not "teach" - selling isn't teaching. Selling is bonding. We also saw it as an opportunity to have them teach us about their industry where they had much more expertise than we did.

Putting it All Together

Once scientists become better communicators and better promoters, we can establish a new stereotype and a new story. Maybe we can become more relatable as a group of professionals, where people will be willing to listen to our advice and judgment because we sound and look just like their friends and family. And maybe if we can focus on telling the story of our science in a more relatable way, with less jargon, and big terminology, we can get the support and funding we need for the amazing things we are discovering every day. Maybe if we focused on the why behind our research and made sure it was a relatable story, the world would trust and accept science more.

Also, as scientists, we are trained to be extremely conservative and to NEVER overstate our findings or claim something that is not 100% true. While being detail-oriented and only stating exact facts is essential for scientific journals and success as a researcher, this type of mentality can inhibit other aspects of growth for scientists. A Ph.D. must wait until they have all the facts and are 100% right. Sales must move forward with 90% failure to search, find and cultivate the 10% who will be a perfect match for what you are selling.

FIGURE 3.1

The PhD and Sales Gap

Success is being 10% right / 9 failures to get 1 breakthrough is also key to science

Scientist	90% success	10% failure
Salesperson	10% success	90% failure

FIGURE 3. THE PH.D. AND SALES GAP

The Ph.D. and Sales Gap - Success is being 10% right - which is also like research 9 failures to get 1 breakthrough is also key to science

If we don't explain the other side of the coin and teach how to communicate and compensate for some of our other strengths, we will keep creating generations of scientists unable to influence and change the status quo. This is the reason Sales for Scientists was written. To be a first of its kind as a self-help guide so scientists, researchers, and engineers can branch out and to build the skills that can change their lives and the world for the better—one story at a time.

The status quo needs to be changed today. And it starts with you, your lab, your team, and your colleagues and friends. Let's change the definition of who scientists are, how we look, how we sound, and most importantly, how we tell stories. Science is the only thing that is going to save our planet from the mounting dooms rising around us. **Now let's start selling!**

CHAPTER 1

A PRIMER ON COMMUNICATION

STORYTIME

"The Birth of the CMOS Imager and the Digital Camera Revolution"

There I was, standing in front of a room full of Taiwanese engineers. I was aware that I might be speaking too fast for their English comprehension, and they might not be catching every word or sentence. Yet, I could see a spark in their eyes, a quickening of interest that was palpable. As I held up a prototype of our CMOS Imager, I saw their curiosity piqued. I passed the sample around, letting them hold it, explore it, and witness first-hand the potential of what we were offering.

The real communication was happening in these moments - not just through words, but through shared experience, touch, observation. I noticed a subtle but vital shift in the room. As I would take the sample away, hands would reach out, wanting to grasp it again, to continue the tactile exploration of this revolutionary technology. It was a clear sign that our message was getting through, not just at an intellectual level but at an emotional one as well.

That day, I learned a valuable lesson about communication - it goes beyond the spoken or written word. It's about engaging your audience on multiple levels, tapping into their curiosity, sparking their imagination, and creating a tangible sense of excitement. This is especially important when dealing with complex, disruptive technology like ours.

Our multimodal communication approach - verbal, visual, and tactile - was key in winning over our international buyers. They didn't just understand the technicalities of the CMOS Imager; they felt its potential, its power, its promise. And that, in the end, was what truly persuaded them to join us on our journey, to embrace the risk and the potential rewards of the digital camera revolution.

This experience underscored the power and importance of communication in sales. Regardless of your audience, your product, or your market, your ability to communicate effectively - to inspire, to persuade, to ignite action - is critical. It's not just about what you say, but how you say it, and how you make your audience feel. Remember, your most effective communication tool isn't just your slide deck or your technical document; it's your passion, your clarity, and your persistence - and sometimes, it's a prototype in the palm of your hand.

WHY YOU SHOULD CARE & WHAT IT UNLOCKS

Communication

If you want to improve every interaction you have with every person, impress others with your knowledge and skills, and influence the world to your liking, then communication is a skill you want to master. To accomplish anything in this world, we need to work with and influence other people through one form of communication or another.

PERSPECTIVES:

A Scientist's Perspective

As scientists, communication is a topic we have seldom devoted our time or attention to. In many cases, we may think the word "communication" is not even relevant to our career. "I don't care about communication, I'm not trying to be a TV reporter. I'm working on an incredible, groundbreaking discovery by myself in my lab." Or maybe you believe there is some importance, but only enough for you to be able to communicate your work to get a grant, get a job, or recruit a student to your lab. But it is not a majority of your focus or effort; at best it's probably an afterthought. This mindset for scientists has set our community back, further widening the gap between our scientific discoveries and the broad dissemination and understanding of our work.

A Salesperson's Perspective

The majority of the scientists I've worked with believe that the value of their work, invention, technology, or product will speak for itself. It should be clear by the facts that this is the penultimate solution for its customers. But in today's world, there is so much noise from other less superior solutions that arrive in email, text, chats, and news, surrounding your customers with noise that distracts them from clear decision-making. While you may have created the perfect solution, it is unlikely the customers will ever know or believe it is without persistent, consistent, and persuasive communication. This might be uncomfortable, and it will be challenging (at least, at first), but if done well, you will be able to make a huge impact and get what you want from your customers and help them be more successful in the process.

QUICK WINS

There are Five Types of communication, as shown in the figure below. Learning each type of communication and how to master it will be crucial for success and growth in your career.

FIGURE 4.1

Types of Communication
and ways to use them

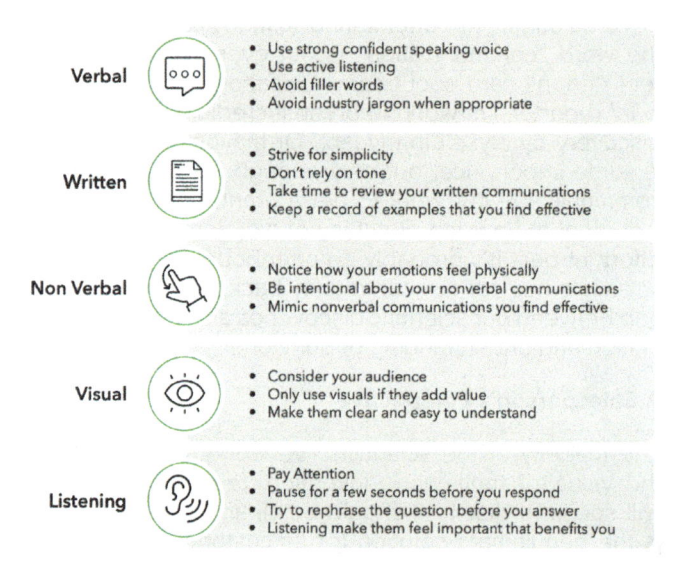

Verbal	• Use strong confident speaking voice • Use active listening • Avoid filler words • Avoid industry jargon when appropriate
Written	• Strive for simplicity • Don't rely on tone • Take time to review your written communications • Keep a record of examples that you find effective
Non Verbal	• Notice how your emotions feel physically • Be intentional about your nonverbal communications • Mimic nonverbal communications you find effective
Visual	• Consider your audience • Only use visuals if they add value • Make them clear and easy to understand
Listening	• Pay Attention • Pause for a few seconds before you respond • Try to rephrase the question before you answer • Listening make them feel important that benefits you

FIGURE 4. FIVE TYPES OF COMMUNICATION OVERVIEW

THE BASICS OF COMMUNICATION

Communication is how we interact with everyone and our world. It is a skill considered simple by many but extremely difficult to master. Without effective communication, we are unable to express ourselves and our ideas to the world, and thus we will be unable to be impactful. However, when communication is mastered, our careers accelerate, our ideas scale, and our lives become healthier and connected to community. Any scientist will not be able to advance in their career if they are unable to communicate their theories and results to get a grant, recruit new students, influence others, start a company, or teach as a professor.

Briefly, there are four types of communication: verbal and non-verbal, written, and visual. Much of the focus of this book will be verbal and visual communication, with less time spent on written and non-verbal. Each type of communication has its strengths and weaknesses, as well as special nuances to master. See below for an overview of each type of communication.

VERBAL COMMUNICATION

Verbal communication is the use of words, both spoken and written, to convey a message or information. It is a fundamental aspect of human interaction and is essential for building and maintaining relationships, and for conveying ideas and information. Verbal communication can take many forms, such as face-to-face conversation, phone calls, video conferencing, and more. It is important to be clear and concise when using verbal communication to effectively convey the intended message and avoid misunderstandings. Effective verbal communication skills are crucial in personal relationships, professional settings, and within a community. Historically, scientists are not the best at verbal communication, failing to tie a cohesive, engaging story that connects with their audience. To simplify teaching scientists to be better communicators, in this book, we focus on *what* is communicated and *why*, so you can choose the right words in a fluid sequence to influence your audience with your message. Mastery of verbal communication can make you feel confident and comfortable in any situation—as a keynote speaker, grant writer, venture capital pitch, and more.

Verbal communication skills are crucial for technical founders who may be selling products or services that require a high level of technical expertise. In these situations, technical founders need to be able to clearly and effectively communicate the value and benefits of their product or service to potential customers.

Here are a few ways in which verbal communication can be especially important for scientists and technologists in sales, their chosen industry, and academia:

1. Explaining technical concepts: As a technical person, you are likely well-versed in the technical aspects of your research, work, product, or service. However, you must be

able to explain these concepts in a way that is easy for non-technical team members and customers or experts in other fields to understand. Using clear, concise language and providing examples can help make technical information more accessible to a broader audience.

2. Answering questions: Customers and your audience will often have questions about the technical aspects of your topic, product, or service, and you need to be able to answer these questions clearly and accurately. This will help to build trust and confidence.

When using verbal communication in sales, it is important to use a strong, confident speaking voice and to actively listen to the needs and concerns of the customer. Avoid using filler words, such as "um" or "ah," as these can disrupt the flow of your message and undermine your credibility, and make others feel less confident in you. Instead, pause to gather your thoughts and then continue speaking. You must also be mindful of industry jargon, as it may not be understood by non-technical experts you speak to. When appropriate, use clear and concise language that is easy for everyone to understand. By being able to clearly and effectively communicate the technical aspects of your product or service, you can build trust and credibility with potential customers and increase your chances of making a sale.

NON-VERBAL COMMUNICATION

Nonverbal communication is conveyed through body language, facial expressions, and other nonverbal cues. Non-verbal communication is an important part of delivering effective messages to our audience. It's *how* we say what we say, the inflections we use, how fast we speak, tone of voice, and pacing.

Body language involves physical behavior, mannerisms, eye contact, and expressions used to communicate without words.

Nonverbal communication is an important aspect of sales, helping build trust and rapport with our audience and potential customers.

Nonverbal communication can convey emotions and intentions that words alone may not. A person with a hunched-over posture, avoiding eye contact while delivering a talk on his breakthrough scientific work will be perceived much differently than a person with great posture, powerful gestures, and strong eye contact.

Scientists often do not put much thought into this, but non-verbal communication tells the audience a tremendous amount of information. Go to any academic conference, put on a noise-canceling headset, and just watch the non-verbal communication throughout every presentation (more often than not, you'll find this is a masterclass on what not to do).

Non-verbal communication can make all the difference with your verbal communication— your words will either resonate with your audience or they won't trust a thing you say. Therefore, communicating non-verbally, it is essential to be mindful of your emotions and how they are expressed physically. For example, crossed arms may indicate you are closed off or defensive, while open posture and eye contact can convey openness and confidence. To be most effective when using nonverbal communication, be intentional about the nonverbal cues you are sending. Pay attention to your body language and facial expressions, and try to use nonverbal cues congruent with your message. If the body language you give off is different than the words you are speaking, it will leave your audience confused and skeptical of your message. It can also be helpful to mimic nonverbal cues that you find effective in other salespeople or presenters.

VISUAL COMMUNICATION

Visual communication are the visual aids we use on presentations, posters, and materials to supplement our verbal and written communication and help to convey our message more clearly. Visual communication is an important tool for scientists and other technologists, as it allows them to simplify complex ideas and make them more accessible to audiences and potential customers. Make sure to keep your visuals clear and easy to understand, and strive for simplicity in your design.

Technical topics, products, or services can often be difficult to understand, especially for non-technical people. By using visual aids, we can help make complex ideas more accessible and easier to understand. For example, a technical founder selling a software solution may use a flowchart to illustrate the steps involved in using the software. This can help customers better understand how the software works and how it can benefit them. Similarly, a scientist can more easily explain a disease pathway using a diagram to show the different stages of the disease and how their work targets one part. Imagine any of the scientific presentations you have given or seen without any visuals! It would be a nightmare! Remember, a picture is sometimes worth a thousand words. Oftentimes, one image or a short video can help us communicate something more clearly than many slides or words. But be careful to only use visuals if they add value—simplicity is best, and your visuals can quickly become confusing and crowded. Other visual aids, such as comparison charts, can help to clearly illustrate the differences between different options, making it easier for your audience to understand how technical topics compare.

WRITTEN COMMUNICATION

Written communication can include emails, letters, reports, and more. Written communication is an important aspect of sales, as it allows scientists and technologists to clearly and concisely convey information to potential customers and stakeholders. One key aspect of effective written communication is the use of brevity. In the fast-paced world we live in, where we may only have a fraction of someone's attention, it is important to be able to quickly and effectively communicate key points of our technical topic, product, or service. By using brevity and concise language, we can ensure the message is clear and easy to understand. This is especially important when communicating with a non-technical audience, as it helps to avoid overwhelming them with technical information. When writing emails, proposals, or other written materials, it is important to strive for simplicity and clarity. Avoid using technical jargon or unnecessarily complex sentence structure, as this will only confuse your audience and make it more difficult for them to understand. Written communication does not convey tone in the same way as spoken communication.

Don't rely on tone to convey meaning or emotion in written communication, as it may be misinterpreted by the reader. Instead, use specific language and examples to convey your message. Every person reads written communication through their own lens and perspective, meaning that the same words have a different meaning to each of us. For example, think about some of the texts on your phone from your family or friends. If a complete stranger read those texts, how different would their meaning be?

Written communication is also often referred back to at a later point, and so the initial contextual information that was once there may no longer be. This is another time we must emphasize being very clear and simple in your language so you don't confuse someone when they refer back to it at a later point. To ensure that your written communication is effective, it is a good idea to take time to review it before sending it out. Check for spelling and grammar errors, and make sure your message is clear and concise. As the expression goes, "If I had more time, I would write a shorter letter."

LISTENING

We need to remember that communication is a two-way process. And to be most successful, we need to practice the other side of communication, listening to others. Believe it or not, in sales, listening may be the most important form of communication. This is an opportunity to better understand the needs and concerns of potential customers. By actively listening, we can modify the sales pitch or our presentation to fit the perspective, mood, or interests of the customer, leading to more successful interactions. A salesperson that doesn't listen is almost certain to fail.

An important distinction needs to be made between hearing and listening. Listening involves actively paying attention and attempting to understand what is being said, while hearing is simply the physical act of receiving sound. We can often have big breakthroughs when we listen because, in a sales presentation or a keynote, we can modify our communication to fit their perspective, mood, or interest. I've been in so many presentations where we are being talked AT about stuff that nobody in the room cares about, and I wished the presenter was listening to our body language or lack of questions. In situations like this, even though it may be awkward, it would be better just to ask your audience what they want to hear about.

Paraphrasing can be a great strategy to improve listening. Paraphrasing is where we repeat back to the customer or audience what they have said in your own words to confirm that you understand their needs and concerns. This also has the benefit of making your customer feel heard and appreciated.

To be a good listener, you must pay attention to body language and nonverbal cues. You also want to be actively asking questions and seeking clarification. This can be a back-and-forth to ensure you fully understand your audience and that they fully understand you. By actively listening and being responsive to the needs and concerns of potential customers, we can increase our chances of making a sale. Listening well makes the customer feel important and significantly increases your ability to close the sale. Make sure to also listen for clues they give you on what they care about and what tone or body language they give off while speaking.

STATS THAT MATTER

- A survey by InsideSales, found that salespeople who use effective communication techniques close nearly twice as many deals as those who do not.
- A study by the National Association of Sales Professionals found that effective communication was the top skill that successful salespeople possess.

SELF-REFLECTION EXERCISES:

Practicing an Elevator Pitch

An elevator pitch is a concise summary of your work, research, product, or service that you can use to quickly and effectively convey its value to others. Practicing your elevator pitch is key because, depending on the situation, you may only have 30 seconds to explain a concept or convince others of the value you're offering. Take something important you are working on in your professional life that is very technical. You will now have to explain this very technical topic and its value to a person with no background in this area—this can be a work or research colleague, a friend, or a family member. Give them your pitch, and afterward, have them explain the technical concept and value back to you. Did they actually understand it? Ask them how your delivery was. Did you speak too fast or use filler words? Practice your elevator pitch a few times, so you can perfect it and feel more confident.

Watching Non-Verbal Communication

For this exercise, there are two components:

1. Watching Yourself:

First, for an upcoming presentation or talk you may have, set up your phone and record yourself giving the presentation. Watch it with and without sound. Are you happy about your body language? Have you noticed any nervous ticks or oddly placed body movements that don't fit with your message? Try delivering the same talk, but now plan out your non-verbal communication, and pay close attention to it.

2. Watching Someone Else:

Watch an accomplished speaker performing their talk. Find a memorable or inspirational famous talk or perhaps a recent TED talk that you loved. This time, you will take notes throughout the talk only on the speaker's body language. What is their body language saying? Do they look confident? Are the big moments in their talk paired with large gestures or other non-verbal communication? After you have taken your notes, watch it again, but this time without any sound.

Ask yourself the same questions about their confidence, their message, and anything else you can glean from their non-verbal communication.

Breaking Down Complexity

For the recent TED talk you watched in the previous exercise, go back through and take note of what visual aids they used. How did they use simple visual aids to communicate very complex topics? What are some of the other visuals they could have used to help convey their message? How many visuals did they use in their entire presentation? What verbal communication was paired with each visual? I'm sure you're surprised at how little was used in explaining so many complex topics.

Now we are going to go through one of your recent slide presentations. Go through each slide and think about the key message. Do you have visuals that help your audience understand that one message? Can the visuals be misinterpreted in any way?

Write a High Impact Summary

One very commonly used verbal communication tool to practice are high impact summaries. These are short, concise summaries that highlight the key benefits and features of a product or service. They may also convey the overall impact or methodology of your research. It can also be a brief summary of a topic with the key messages included.

By using high-impact summaries, we can quickly and effectively convey the value of their research, product, or service to the world.

You will now write your very own high-impact summary. Choose a product or service to sell and practice crafting a high-impact summary highlighting its key benefits and features and why someone would want to buy it. Practice using clear and concise language to convey the key points of the product or service and pay attention to the tone and style of your writing.

Try three versions

1. Your first attempt

2. One with only 3 paragraphs

3. One with only 3 bullet points

Sales for Scientists

COMMON MISTAKES

Below we outline some of the most common and costly mistakes with each type of communication. If you make enough of these mistakes, your audience may come away confused and unhappy, and won't make the sale.

Verbal

- Using too many words, overexplaining, and lack of brevity means your audience will have to pay more attention for longer and give them more opportunities to consider why what you're saying isn't right for them.
- You may think jargon will make you sound smart, but all it does is cause your audience to lose interest.
- Filler words like "um" or "ah" are one of the most common mistakes people make. Strict practice and focus are required to filter out "ums" and "ahs." The more filler words used, the less confident you seem to your audience.
- Talking too fast or too slow. You need to think about your pacing when you speak.
- Talking monotone with no enthusiasm, passion, or excitement. If you aren't excited, your customers won't be either.

Non-verbal

- Misalignment or lack of gestures and actions with verbal communication can leave the audience confused. Make sure your timing with gestures and body language is planned and natural.
- Stiff body posture, awkward hand gestures, or other nervous twitches that may cause your audience fear or discomfort.

Visual

- More visual aids are not always better. Far too often, new speakers will use a surplus of visual aids without

any structure or natural guidance for the audience to follow.
- Use of too many colors or fonts can leave the audience searching for meaning where there may be none.

Written

- Complexity is never better, and good writing is as simple as possible.
- Take for granted that your message can be read and understood without full context or tone. This is rarely the case and you don't know when your email will be read and by who!

Listening

- Hearing and listening are not the same. Make sure to be active and take mental or written notes in conversations or during Q&A.
- Often people will stop listening once you finish speaking, and think it is a chance to turn off the brain. Wrong – this is when you need to be even more engaged.
- Trying to multitask is a recipe for disaster and should never be done, especially in important situations and during conversations.

SUMMARY

Communication is a foundational skill for sales and a skill everyone can learn and benefit from. Yet, there is a big gap in education on this among the scientific community. Build a strong foundation starting with this chapter, practice the exercises and skills, and turn on the mindset that communication is a lifelong skill that can be practiced every day. Every interaction you have with others is a chance to practice these skills.

CHAPTER 2

PERSPECTIVE - THE SECRET SKILL TO SELLING SUCCESS

STORYTIME

"The Lifetime of Grants vs. Commercial Reality"

After a decade of launching pioneering, novel technologies into mobile, consumer, and enterprise sectors, I was called upon by a few familiar faces. Early founders of a company I previously worked with had transitioned to an MIT spin-out company and saw me as the solution to a challenge they faced. This new venture was the brainchild of three Ph.D's, with a fourth added to the mix as the CEO. Two were actively engaged MIT professors, another was a fully committed graduate student, and the CEO was a former technical lead at IBM's research center, having steered the victorious Deep Blue Chess Project against world champion Garry Kasparov.

Each one of these individuals was a beacon of brilliance in their respective fields, wielding considerable influence. However, this was their maiden voyage into the sea of commerce. As a young professional, just into my thirties, I was equipped with almost a decade of global experience in launching innovative technologies.

The business had been in motion for three years, buoyed by early funding from strategic partners who were part of the founders' "benefactors". Success and funding came in the style of academic "grants": pose a hypothesis, formulate a proposal outlining the problems that the project could solve, and await the benefactor's goodwill to fund the high-risk venture or research.

Our discussions often grew heated, revolving around business strategies, securing deals, and understanding customer needs. It took me two long years to fully grasp the essential gap in our perspectives. The founders had flourished in the realm of academia, mastering the art of obtaining grants, and their belief system had evolved from this foundation. They lacked a business perspective – the viewpoint of a product manager, an engineering VP, or even the CEO of our customers. They were missing a perspective about success strategies, growth necessities, and an understanding of what our customers needed to make decisions that would bolster our business.

Their academic careers had flourished, guided by their finesse in navigating the grant and publishing processes. They had the blueprint to devise pitches and secure funding for their cutting-edge research.

However, the commercial world was foreign to them. They had never worked in a commercial business or made operational decisions on crucial aspects like profit and loss. In commercial environments, where short-term focus rules, the grant process falls short. Customers think about immediate, rudimentary problems: "What's in it for me right now?" They crave the assurance of immediate commercial and personal gains, the security of a stronger, richer, and more competitive company. A fleeting overview of the technology suffices – perhaps 10-15 seconds, especially if complemented by some appealing marketing jargon. To make a mark, you need to grasp and adapt to different perspectives and be conscious of their potential impact on your business. The market's perspective drives growth, velocity, and shareholder value. Our technology-first perspective rendered us tone-deaf to the needs and expectations of our partners, customers, and, ultimately, our acquirer. Although we celebrated a successful exit, the value realized was a meager 10% of our primary competitor because they evaluated us solely on our technology's merit rather than our impact in the market.

WHY YOU SHOULD CARE AND WHAT IT UNLOCKS

Perspective

When you master the art of understanding perspectives:

- You're able to form deeper connections with everyone you interact with
- Your ideas have more influence and impact, allowing you to increase the virality of your message, making it easier for you to win more arguments and decisions.

Whether you go into sales, industry, academia, or become a stay-at-home partner, understanding perspective allows you to deal with people more effectively.

PERSPECTIVES:

A Scientist's Perspective

A scientist's perspective is a rarely considered concept. We are:

1. Deep experts in our field and our work, often being the single most knowledgeable person on a topic
2. Surrounded almost constantly by peers of similar and high-technical levels. When we regularly and almost exclusively engage with peers like this, we never have to recraft our perspective and messaging to communicate clearly to others. Or when we do, it is only a minor adjustment.

3. We often will work in a silo, focusing on every detail of our work, knowing it all is crucial for the outcome we hope to achieve.

These forces are working against us, and why we as scientists need to try even harder to work on our perspective skills. We are rarely confronted with the downsides of our failing to consider other perspectives, and so we go further and further until we find out how big of an issue our lack of perspective is.

A Salesperson's Perspective

The most successful salespeople know that true success comes from understanding and connecting with the customer's perspective.

Early in my career, I took sales training classes called "Strategic Selling" from Miller Heiman. Even as engineers and scientists, it is important to understand and incorporate strategic selling into your work.

After more than a thousand sales meetings with deep tech founders and leaders, I've observed that the most common flaw in their approach is a narrow perspective, only focused on what they have and what they want from the customer. In an ideal world, they would never need to talk to a customer; their product or service is just so good that the customer would wire the money in advance of a sale. The second-best scenario is they would walk into a room with the senior executives of the customer, present a 37-page dissertation on the technical merits of ant saliva for machine lubrication, and the customers would bow down and throw money at them.

The frank reality, as discussed in venture capitalist circles, is that "nobody cares about you, your product, or your discovery." Sure, the customer may be meeting with you; they may invite others to see what you have to say, and they might even join you for lunch, but they don't care about you. They care about themselves. To get what we want, we need to know what the customer cares about…their perspective.

STATS THAT MATTER

- Customers who feel that a company understands their needs are more likely to make a purchase. According to a survey by PwC, 73% of consumers say they are more likely to make a purchase from a company they feel understands their needs.

- Personalized experiences lead to higher customer satisfaction and loyalty. A study by Epsilon found that personalized experiences lead to a 19% increase in customer satisfaction and a 20% increase in customer loyalty.

- According to a survey by the White House Office of Consumer Affairs, 88% of consumers are more likely to make a purchase from a company they feel understands their perspective and needs.

QUICK WINS

You Have a Unique Perspective

The first step to understanding others' perspectives is to first dig deep and understand your own. You have a unique LENS in which you view the world as well as your own BIASES, both of which you need to understand.

Ask "Why" to Truly Understand Other Perspectives

To understand another person's perspective, so we can succeed in sales, we must dig deep and ask "why." Not just – do they care about X, Y, or Z, but *WHY* do they care about X, Y, or Z?

DIGGING DEEPER UNDERSTANDING YOU

Your Lens

Every experience you have had in your life, your highs and lows, education, accomplishments, failures, and biggest lessons learned, all create your unique perspective. Imagine now that the research project you have been working on for five years, is suddenly handed over to a stranger, who is expected to understand all the intricacies and make the next big decisions. I bet you would feel extremely scared and worried, as this person lacks the perspective to handle the situation. What would be obvious for you in this situation would be pretty complicated for a stranger. Use this analogy on a much grander scale and apply it to your life. Pull out the key experiences and foundations that shape you and your personal and professional decisions. As we more clearly understand our perspective, we can also use this to better understand the goals we have in life, both big and small.

Your Bias

As we gain insight into our perspective, we will also be able to identify certain biases we may have that skew how we perceive and how we communicate in every situation. Bias is a systematic deviation from an objective perspective, affecting our decisions, judgments, and actions. In sales, and as scientists, we need to be extra careful how biases impact our decisions and relationships, and how we consider our perspective compared to others. Understanding our biases can help us overcome them, while understanding others can give us a better chance at understanding their perspective to positively influence them. We have both conscious and unconscious biases, our unconscious ones being more difficult to manage and change. Maybe we are impatient because we have been burned too many times after waiting too long for a colleague's buy-in. Maybe we are scared to accept long research and development timelines because we have been burned in the past. Whatever the bias may be, however minor or major, visibility and acknowledgment are the only ways we can get past them and prevent them from impacting our work.

Some of the more common types of bias to be aware of and consider:

1. **Cognitive bias** occurs when our brains use mental shortcuts (heuristics) to process information and make decisions. Often, these mental shortcuts can lead to biased judgments as they rely on an incomplete set of information. If we fail to acknowledge our cognitive biases when selling, it can lead to a narrow-minded approach to selling to our customer with less success. Some examples of cognitive biases are:

a. Confirmation bias: tendency to seek out and give more weight to information that supports our existing beliefs, while ignoring information that challenges our beliefs. You may experience this bias when you are trying to sell your idea to friends and advisors, and avoid presenting and seeking feedback from those that disagree or poke flaws in your concepts. In reality, these are the people you want to engage more with and try to convince. When first learning to sell or learning to sell something new, you will have to navigate this bias as your audience may subconsciously avoid your information and ideas because it goes against their long-term beliefs.

b. Sunk cost bias: continuing to invest effort, time, and money into something because we have already invested a lot into it. This bias is strong, especially when you have spent years on your technology for your startup or research. One of the biggest lessons with this bias is to fall in love with the problem you are solving, not *how* you are solving it. This is also a bias you will have to overcome selling to businesses, as individuals that have spent years getting a system or technology to work at a company, will feel a bias to not change their technology, even if yours is much better.

2. **Social bias** occurs when our perceptions, attitudes, and behaviors are influenced by the social groups to which we belong or the social norms and expectations that exist in our society. First, recognize if any of these unwanted biases exist, and then work to eliminate them by simply keeping them top of mind. If you do not, it can negatively impact how you sell to a broad group of people. Some common examples of social biases include:

   a. Stereotype bias: tendency to make assumptions about people based on their membership in a particular social group.

   b. In-group bias: tendency to favor or show more loyalty to people who belong to our own social group.

3. **Emotional bias** occurs when our emotions influence our judgment and decision-making. Emotional biases can be

positive or negative, depending if the use of a product, for example, provided a feeling of resentment or joy in the past. Emotional biases are often unconscious, making them difficult to control. Emotional bias can be important to navigate when selling new technology to a company, as many may feel a tremendous amount of loyalty and happiness or disappointment from a product or service. If you are not able to overcome emotional bias, it may be a barrier to selling your product effectively.

4. **Technical or Scientific bias** is the tendency for technical experts, such as engineers and scientists, to focus on the technical aspects of their product or service when selling. This bias also occurs when a person's knowledge or expertise in a particular technical field influences their judgment or decision-making. As scientists and entrepreneurs, we must be aware of this bias because it may influence the solution we develop or the technology we choose, only because we are comfortable with it and used to it. Yet again, the sage advice of falling in love with the problem, not the solution comes in to play here. When selling, this can be a problem because the customer may not have the same level of technical knowledge and may not be interested in the technical details. This is a common bias for scientists because we assume everyone around us has the same technical depth in our hyper-focused area. To avoid technical bias, understand the customer's full perspective and educational background, as well as language commonly used to make it more accessible and relevant to the customer. When selling our product, we need to recognize our technical bias and adjust the level of technical detail accordingly to each stakeholder or audience we are presenting to.

UNDERSTANDING EVERYONE ELSE

Understanding our own perspective is a challenge that we may never fully understand, yet now we will attempt to understand another person's perspective—a massive chasm to cross. When we communicate our ideas to others, we are attempting to cross this chasm and understand the perspectives of our audience. Mastering and considering others' perspectives may be the single most important skill one can ever learn. In the same way that we evaluate our own perspective, we must do with every person we interact with.

Don't get too uncomfortable; we already do this to some extent daily with everyone we interact with. When you talk to your family who may not know about your work, you consider this and explain new topics to them or provide only information relevant to them. When you meet someone new, you will often ask questions to better understand who they are and their background, so you can better frame your conversation together. We have a good base instinctively, but to get to the next level of understanding others' perspectives, we will need to go a little deeper. Think of the first layer of understanding someone's perspective as more objective fact-based questions, such as their education, job history, age, and other details we covered in looking at our perspective. The deeper layer we need to explore covers the wants, desires, and needs of our audience's perspective. "*Why*" can be one of the most important questions we ask when looking into perspective, which will help us better influence others. The end goal of our selling and interacting with others is to influence them to take action. If we are to do this, we must really consider what motivates them, what they need, why they may need it, why they may feel a certain way, why they care about your presentation, and why they would choose to take a certain action. Before any important interaction, whether it be a presentation, a sales call, or even a date, we must consider the other person's perspective in depth.

PERSPECTIVES IN SELLING

In the world of sales, it is easy to get caught up in the numbers and the constant pressure to hit targets and quotas. However, the most successful salespeople know that true success comes from understanding and connecting with the customer's perspective.

When selling from a perspective, the focus is not on the product or service itself but on the customer and their needs. This means taking the time to truly listen to the customer, understand their pain points and challenges, and tailor your sales pitch to address their specific needs.

To be a successful selling scientist, it is crucial to internalize this perspective and let it work deep into your bones. This is about emphasizing the importance of understanding the customer's perspective. To get what you want, you need to let go of your ego and focus on the customer's needs.

One effective way to do this is by using the "6 P's" approach to selling, which involves understanding the customer's overall purpose, identifying their problems, providing proof, showing possibilities, developing a customized plan, and discussing pricing. By showing the customer that you truly understand their perspective, you can build trust and rapport and ultimately lead to a successful sale

ONE MESSAGE ACROSS MANY PERSPECTIVES

To illustrate how important perspective is to consider, here we demonstrate how one discovery from a scientific researcher gets communicated to the many different perspectives across a large organization.

Situation: A breakthrough from your research has allowed a material used in the phone XYZ Company sells to require 25% less material but maintain the same battery life.

Scientist

- **Perspective**: Every amazing detail of what I've just done is so important and amazing that everyone I speak with will want to know. They will want to understand how the breakthrough came about, what I did to achieve this, and why it is such a breakthrough.
- **Message**: By manipulating the molecular composition, as well as the ...detail.... of the adhesion process in the ... detail..., we are able to pause the crystallization process... detail ... Detail... detail... detail... And yes as a result of all this, there could be less material required to maintain the same material properties. Under rare ABC circumstances, we are able to reduce the material required by 25%.

Now this scientist will need to convince the other stakeholders in the organization of the importance of their work and why they should devote their time and resources to make it happen. Let's see the other perspectives of stakeholders in the organization, what they care about, and how the scientist may be able to sell to them, considering their perspective.

Research and Development "R&D" Lead at XYZ Company

- **Perspective**: I'm working on dozens or hundreds of important research topics. I need several of these to have breakthroughs like these per year. If I don't, I may lose my role. This breakthrough may help me to get the next promotion I'm looking for.

- **Message**: A scientist may sell their idea to an R&D lead by highlighting the research breakthroughs and the big impact it can have on the product lines that research and development contribute to. The R&D lead is held accountable for their research breakthroughs being connected to end product goals that can then deliver increased sales goals. While the R&D lead will be excited about the brilliance and the technical details, the scientist will need to clearly connect to why this matters for the product line or impacts costs. For example, if the scientist was to solely focus on the amazing research breakthrough, but is unclear why this has meaningful benefits, the R&D lead will not care too much, as it is not how they are measured in terms of success. If it is an incremental breakthrough from the scientist, they can focus on the path to a longer-term goal that impacts product lines and manufacturing margins.

Production Manager

- **Perspective**: I need to make the production line move as efficiently as possible, and am willing to look at new optimization methods, but only if they will not impact my production negatively. I am held responsible for shutdowns even if they are not my fault.

- **Message**: When selling to a production manager, a scientist should focus on the practical applications and benefits of their product or service to the production process. The production manager is likely to be more interested in how the product can improve efficiency, reduce downtime, and enhance the quality of the final product. A scientist may sell their idea to a production manager by highlighting the potential benefits it could have for the company's production processes. This could include new technologies or processes that could improve the efficiency, quality, or cost-effectiveness of the company's production operations. The scientist could also discuss how their idea aligns with the company's production goals and how it could help the company achieve its production objectives. For example, the scientist could start by discussing the purpose of the product or service and how it can help the production process achieve its overall goals. They could then identify the problems that the production process is facing, such as inefficiencies or quality issues, and how the product can help solve them. Next, the scientist could provide proof of how similar breakthroughs have helped their company solve similar problems. The scientist could then develop a customized plan that outlines how the product will be integrated into the production process and how it will help the company achieve its goals. By using this approach, the scientist can show the production manager that they understand their needs and can provide a solution that will help improve the production process.

Product Manager

- **Perspective**: We need to deliver new experiences for our users that will excite them and make them continue to choose our phone line. I want to work with R&D and the technical team to deliver innovations that matter to our customers. I need to find a way to communicate technical breakthroughs into experiences for our users.

- **Message**: A scientist may sell their idea to a product manager by highlighting the potential benefits it could have for the development of a new product. This could include new technologies or processes that could improve the product's performance, design, or cost-efficiency. The scientist could also discuss how their idea aligns with the company's product roadmap and how it could help the company achieve its product development goals. For example, a scientist could focus on the fact that users care deeply about the phone's weight and the ability to carry it with ease in their hands for long periods of time. We partnered with our R&D and technical teams to find a way to reduce our phones by 25%, which weigh less than all competitors. This would be compelling for a product manager to devote their time and energy and to then have the product manager help you convince and influence others.

Marketing Manager

- **Perspective**: We need something unique and differentiated that will excite and entice our users. A new technology we can brand.

- **Message**: A scientist may sell their idea to a marketing manager by highlighting the potential benefits it could have for the company's marketing efforts. This could include new technologies or processes that could improve the performance of the company's marketing campaigns or the effectiveness of its marketing materials. The scientist could also discuss how their idea aligns with the company's marketing goals and how it could help the company achieve its marketing objectives. The scientist could provide data or evidence to support their claims and show why their idea is worth investing in.

VP of Engineering or Manufacturing

- **Perspective**: We need high volume and stability in our product line. New technology can be an advantage, but proven yield, process, and efficiency are required to meet our high-volume production demands. Only vetted technologies will touch our production, as there are a series of certifications and risks that could be catastrophic if gone wrong.

- **Message**: A scientist could sell their idea as a new application of our existing materials. They can focus on how it is low risk and will seamlessly fit into our existing supply chain without changing the manufacturing process. A scientist may sell their idea to a VP of Engineering by highlighting the potential technological advancements it could bring. This could include new materials, processes, or designs that could improve the performance, reliability, or cost-efficiency of the company's products. The scientist could also discuss how their idea aligns with the company's engineering and manufacturing goals and how it could help the company achieve its engineering objectives.

CFO

- **Perspective**: We need to maintain our profit and loss and cut any projects that are too costly without return. We are not currently investing in projects that have a clear return on investment and a high likelihood of success because we can't afford many failures at this point.

- **Message**: A scientist may sell their idea to a CFO by presenting the potential financial benefits of their research or project. This could include potential cost savings, revenue opportunities, or other financial advantages. A CFO will care about some of the key financial health metrics that they monitor and maintain for the company. The CFO will also be aware of any current upcoming cost-cutting or spending focus areas or themes. If the scientist can understand some of the upcoming spending areas or biggest cost areas the CFO is focusing on, the scientist can position their message to fit into these areas. A scientist that plans and pitches with this strategy will be more successful than not. For example, a scientist can speak to the long-term sustainable advantage in terms of cost due to their innovation. As a result, the company's manufacturing costs will drop substantially, which means they have flexibility in price to help outcompete others in the market.

CEO

- **Perspective**: We need to invest in groundbreaking technologies and ideas that can sustainably create competitive advantage. We are in a highly competitive market, our stock price is falling, and my job will be on the line if we don't turn this ship around. Anything we pursue will also need to drive our company mission and vision, and set us up for the next several years of success.

- **Message**: A scientist may sell their idea to a CEO by highlighting the potential impact on the company's bottom line. This could include potential cost savings, increased productivity, or new revenue streams. The scientist could also discuss how their idea aligns with the company's goals and mission and how it could help the company gain a competitive advantage in the market. A scientist should focus on the value and benefits of their product or service to the company, rather than the technical details. The CEO may not have a deep understanding of the technical aspects of the product and may be more interested in how it can help the company. They could then identify the problems the company is facing and how the product can help solve them.

FIGURE 5.1

Perspective Matrix
The level of technical information & detail provided is reduced ...

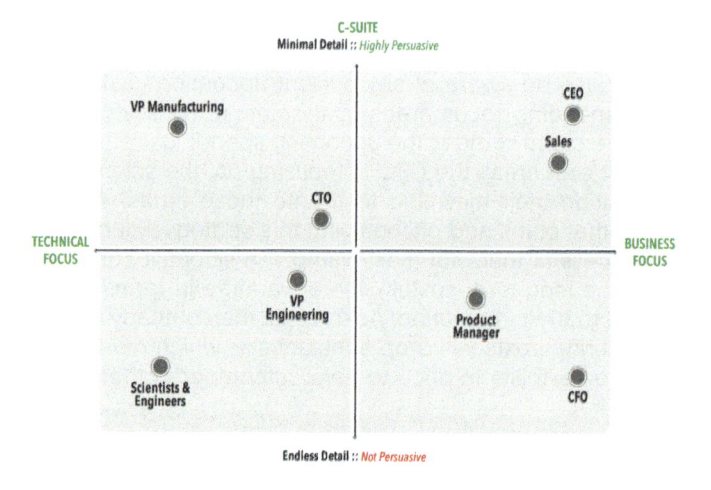

Figure 5. Chart of Information to Provide by Role

(the above figure is meant to be illustrative rather than definitive and varies by organization)

SELF-REFLECTION EXERCISES:

Understanding Your Lens & Others'

In this exercise, we will spend some time digging deeper and reflecting on what makes your perspective unique. To start, pick another person at work or maybe a friend, and compare your perspective to theirs. Create a table answering questions about each of your unique perspectives. Here are just a few examples to help you get started:

- Education
- Birth location
- Age
- Career history
- Favorites
- Religion
- What is most important to each of you
- Family size; siblings, cousins, etc.

Now that you have an idea of varying perspectives, you will craft three situations to help you evaluate how our perspectives can vastly impact each situation. Think of three situations you were recently in where you were dissatisfied, frustrated, or angry. Write down why you felt this way and how you think your unique perspective may have influenced you to feel that way. For the second part of this exercise, imagine the other person in those situations. Do they believe or end up feeling the same way as you? Are they angry, or are they perhaps indifferent or happy? It is a powerful lesson to see how others respond to the same situation, with the only difference being perspective.

The Johari Window

The Johari window is a tool initially created by psychologists to understand conscious and unconscious biases and the perspectives of oneself and others in order to enhance communication. The Johari window can help us to identify blind spots or areas of ourselves that we keep hidden from others. This will help you realize there is a lot below the surface for every person we interact with and try to sell to. Below is a figure with four quadrants you will fill out as part of this exercise. First, pick a person or a group at work to focus on. Next, you will fill out each quadrant with your attitudes, behaviors, emotions, feelings, skills, and views.

- **Quadrant 1:** list the above items known to yourself and others. This is where most communication occurs in your world, and where both you and others are aware of your perspective. Here you can recognize your perspective that is visible to others, and that everyone else you interact with only has a portion of their perspective for others to see.

- **Quadrant 2:** list the above items not known to yourself but to others. These are your blind spots. This quadrant will most likely require speaking with others in the identified group to see what lies there. These can be some of your individual biases or other new unique aspects you were not aware of.

- **Quadrant 3:** list the above items known to you, but you keep from others. This will be your façade quadrant. Things like fears and secrets are often what will be found here. The more trust and openness you have with the world and others, the more you can be open about this quadrant and make it smaller. If you are able to access this layer with customers, it is a good sign of trust and that you can really solve deeper problems if need be.

- **Quadrant 4:** this may be very difficult or impossible, but here we list items that are both unknown to ourselves and to others. This may be past items we discovered, or it may require a lifelong pursuit to identify the items in this area. Regardless, we must recognize that there are items that exist in this quadrant. Open communication is the best chance at uncovering these.

FIGURE 6.1

Johari Window

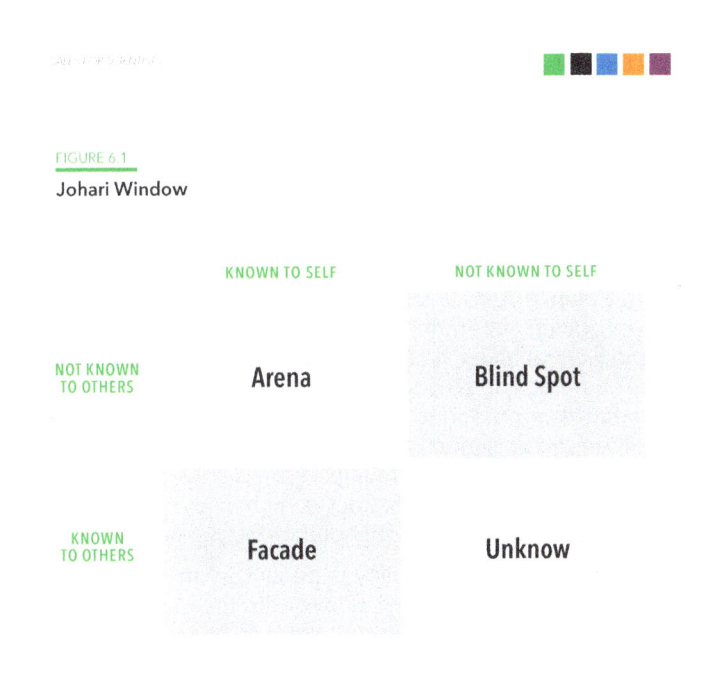

	KNOWN TO SELF	NOT KNOWN TO SELF
NOT KNOWN TO OTHERS	Arena	Blind Spot
KNOWN TO OTHERS	Facade	Unknow

Figure 6. Johari Window

COMMON MISTAKES

- Lack of Empathy - For many scientists, our heads stay in objective facts and data, and when we interact with others, we stay on this level. However, we cannot forget to dig a layer deeper and consider more subjective aspects to fully grasp another person's perspective.

- Failure to Consider Technical Depth - Many scientists assume everyone around them has as high a technical understanding as they do, or if they don't, then they should. "Why should I dumb down my talk or explanation?" However, this will lead to your message being missed by the audience. Always match the audience's technical perspective, and if you aren't sure, it is better to err on the side of caution.

- Not Using Your Customer's Lingo - Make sure to always use clear language that your specific customer understands and uses regularly. If they use an abbreviation common to them, you better know it. Often, you can quickly out yourself as not an expert using the wrong language.

- No Balance Across Multiple Perspectives – It's likely that we regularly interact with a group or mix of many people and perspectives. It can be challenging to accommodate all of these, and so often scientists may only pick one perspective or person to speak to. This is a mistake, as you will be missing much of your audience. Scientists should find balance between multiple perspectives and technical levels, and then abstract a higher-level perspective that can match the majority of their audience.

- Being Inflexible - Flexibility is necessary in sales. Technical people very often fall in love with their idea and work and are scared or refuse to taint it in any way; their perspective *has* to be right. Be flexible with your idea and your perspective as it will reach a larger audience.

EXAMPLE: BEFORE / AFTER

Situation: A salesperson is meeting with a decision maker at XYZ company to consider buying their new technology widget

- Before: The salesperson does not understand the customer's perspective, the value the widget brings to them, and why they are considering buying the product. The salesperson focuses on describing all the features of the product, the technical advancements, and the value it could bring to XYZ company. The salesperson talks nearly the entire time. As a result, not much time is spent around the customer's specific needs and situation, and the salesperson does not learn what are the key problems the customer is facing. The customer says they will think about it and get back to them.

- After: The salesperson does further research on the customer's situation, speaks with several stakeholders, and considers the impact on the current economic environment for the customer. The salesperson understands that labor issues are the top concern and other concerns are secondary. The salesperson introduces the widget but then allows the customer to speak on top-of-mind problems, as well as those that could be relevant to the product. The salesperson also now learns that the widget solves a portion of their problem relating to labor and that the main benefits of the widget solve secondary problems. After understanding the customer's perspective, the salesperson positions the widget to focus on addressing labor issues and several other benefits. The customer feels that their perspective and needs are heard and decides to move forward with a pilot program geared around key performance indicators (KPIs) that measure labor performance.

SUMMARY

As a scientist, salesperson, and human being, we must recognize and take into account that everyone, including ourselves, views the world through a different lens; we all have a unique perspective. The sooner we recognize that and work with it rather than against it, the sooner we can have a greater impact on the world. More sales, better presentations, bigger awards, healthier and more productive interactions with the people we come into contact with, and more.

CHAPTER 3

CREDIBILITY - THE TECHNICAL HAMMER

STORYTIME

"Showcasing our technical knowledge, at the right time to close the deal."

This story has become foundational in how I approach sales and was told to me by my mentor, who is the best selling scientist I know. In the early days of the digital imaging industry, Olympus was considering investing $3 million in a Boston start-up to commercialize the world's first CMOS imager-based digital camera systems. Today, billions of CMOS imagers are shipped every year in almost every mobile phone, car, and doorbell, but in the 90s, they were all lab-based and had significant technical potential to change the world, if we could only overcome the technical challenges and get more than two to be made using the same process. There were significant credibility challenges to overcome for both the team and the technology.

The technical potential was revolutionary, but the technology was raw, risky, and unknown. Olympus sent a team of Ph.D's with hundreds of questions to Boston to meet with the start-up's founding team. There were only eight people in the company (half of whom were Ph.Ds, including my mentor), but they were the world's leading digital imaging experts. For three days, the Olympus team met, asked questions, and ran scenarios to model the risk of funding the CMOS imager market and test the credibility of the startup team. For three days, the start-up answered the questions in detail with 20-30-page reports and details about the approach and solutions. It was a Ph.D. battle royale; a technical slugfest trying to model the unknown.

However, nobody could really know the answers until the work was actually done. Experiments were run, and both the science and transition to a manufacturable product were completed. My mentor finally said, "Enough. I believe the real question is: are we smart and creative enough to make this solution work?" We can go through scenarios, but we have to make a go/no go decision. Here's my proposal: the Olympus team should meet overnight to come up with ONE question. It can be any question you like, but only one question. If we answer it, you'll fund us; if not, we'll go our separate ways." This is the birth of a process we called "the Technical Hammer."

The end of the story is that, yes, Olympus came back with one question. Yes, my mentor's team answered the question, and the world's first CMOS-based digital still camera was created. This led to the first consumer digital camera with Olympus, the first cellphone camera with SK Telecom in Korea, and the first CMOS automotive camera with Daimler. This accelerated the future toward the world's first CMOS mobile and automotive projects, which are the foundation of everything from TikTok to self-driving cars. The creation of all of this, all dependent on establishing credibility at the exact right moment with the audience.

Long story short, learn to harness the power of the technical hammer. Wait for the crucial moment to drive the nail home with full force, establishing your credibility and paving the way for success.

WHY YOU SHOULD CARE AND WHAT IT UNLOCKS

You need to care about credibility as a scientist because it can have a significant impact on the success of your efforts in nearly every sales situation or presentation. When you establish yourself as a credible and trusted source, you are more likely to build trust with your audience, your peers, and your potential customers or clients, which can make them more receptive to your ideas or products. This can ultimately lead to increased sales and success in your endeavors. Credibility will also help to differentiate you from competitors and increase the perceived value of your products or ideas. By establishing a reputation for expertise and credibility, a scientist can make their products or ideas more attractive to potential customers or clients. Finally, credibility is essential for building and maintaining strong professional relationships. By establishing yourself as a credible and trustworthy partner, you can open up new opportunities for collaboration and partnership, which can be critical to your long-term success. Credibility alone is often the difference between success and failure.

PERSPECTIVES:

A Scientist's Perspective

As scientists, credibility only comes into mind in the context of journals, publications, and the data within the research. "Was the work done here credible and up to a standard that we can trust their outcomes to build off of for our work?" Credibility is especially valued in the scientific community because of the nature of the exchange of information and the advancement of knowledge. But in the business world and beyond, credibility should have a much larger meaning and context for scientists to be successful. To gain trust, be engaging in our storytelling, and to close deals, we must build credibility with our audience. If a scientist presents their work on stage, but the audience does not view them as a credible scientist, then that scientist and their work will have zero impact. Credibility should be a steppingstone and part of a scientist's path to success via sales and career goals.

A Salesperson's Perspective

As a deep-tech sales partner to a founding Ph.D., I have had the opportunity to sell disruptive technology to a number of high-level executives, including four billionaire CEO tech company founders. These executives were running multi-billion-dollar businesses and had access to a wealth of engineers, scientists, and Ph.D.s, but what they truly wanted from me and my technical partners was a glimpse into the future. They knew that the bleeding edge of technology came with risk, but they needed someone to be their crystal ball.

In order to drive credibility for the future, it is important to separate the messages. As the sales partner, I can speak confidently to the company, technology, and market, while the scientists can provide rock-solid, 100% correct statements of clarity. However, avoid using "disclaimers" as they can dilute your credibility with decision-makers. Instead, it is important to be clear and direct with key points. Focus on the future and provide confidence on the likelihood of success. While the executives are looking for 0.014% more efficiency from mature processes, when speaking with a scientist, they want to "know" what can and will happen, either to identify a disruptive opportunity or a risk to their business.

Secret Tip for Scientists: Whiteboarding: One of the best tools for building credibility is to whiteboard the information with their technical lead in front of the executive team. This allows the executive and the larger group of supporters to see the expert's thought process and understand the technical details clearly and concisely. By doing this, they are not only judging the technical merits that they most likely won't understand but the relationship and ability to partner with their team to make a high-risk bet more likely to succeed.

STATS THAT MATTER

- According to a survey conducted by the *Nielson Global Trust in Advertising* report, 83% of respondents said they trusted the recommendations of friends and family, while 66% said they trusted consumer opinions posted online. This suggests that people are more likely to trust and believe in a product or idea if it is endorsed by someone they know and trust, or by a large number of people online.

- A study published in the *Journal of Marketing Research* found that credibility is a key factor in determining the perceived value of a product or service. In this study, researchers found that people were willing to pay more for a product if it was endorsed by a credible source, and they were more likely to recommend the product to others.
- According to a study published in the *Journal of Consumer Psychology*, people are more likely to trust and believe in a product or idea if it is presented by a credible source. In this study, researchers found that people were more likely to purchase a product if it was endorsed by an expert or someone with a strong reputation in their field.

QUICK WINS

Establish Yourself as a Thought Leader

Attending and presenting at conferences, creating content in a variety of ways, posting your ideas on LinkedIn and social media, and guest blogging all help to establish yourself as a thought leader. The more content to point to and the more people who believe in it, the more credibility you establish, and the easier it will be to convince more people of your ideas.

Build Strong Relationships for Warm Intros to Cold Prospects

Naturally, we all have a barrier before accepting others' ideas about the products and services we "should" buy. But when the idea is presented by someone we trust, transferred credibility breaks through that barrier. As salespeople, we need to leverage warm introductions from those we have built relationships with to increase our odds of convincing others' that our idea is a winner, thus boosting our credibility and, in turn, our sales.

Use a Balance of Humility, Honest, and Confidence

Honesty and humility are two vital ingredients to help establish yourself as a credible source. Honesty is always essential; being caught in a lie of any size even one time can destroy years' worth of relationships and effort. Your humility allows them to open their door for you to influence them. It's vulnerability at its best. Confidence helps close the deal.

DIGGING DEEPER

Credibility is a critical element in any sales situation, but even more so for scientists who are seeking to sell their ideas or products. In the world of science, credibility is closely tied to the concept of expertise, and people are more likely to trust and believe in a scientist who is seen as an expert in their field.

Let's start this section with a scenario to highlight why scientists need to learn how to build and leverage credibility.

Imagine a scientist who has developed a new type of solar panel that is more efficient and durable than anything currently on the market. This scientist is seeking to sell their solar panels to business leaders, such as CEOs and product managers, in order to bring their technology to a wider market. In this scenario, the scientist's credibility would be a primary factor in the success of their sales efforts. Business leaders are likely to be highly skeptical of any new product or technology, and they will want to be sure that they are investing in something reliable that will provide a good return on their investment. To build credibility with these business leaders, the scientist will need to take several steps.

First, they will need to clearly demonstrate their expertise in the field of solar panel technology. This could involve presenting data on the performance of their panels, discussing their unique features and benefits, and explaining how their panels compare to their competitors'.

Second, the scientist will need to establish their credibility as a trustworthy and reliable partner. This could involve sharing information about their research and development processes, discussing their past successes and any awards or recognition they have received, and being transparent about any potential risks or limitations of their technology.

Finally, the scientist will need to be able to clearly communicate the value of their solar panels to business leaders. This could involve presenting data on the cost savings and environmental benefits of using their panels, discussing potential partnerships or collaborations with other companies, and highlighting any additional features or benefits their panels offer.

By taking these steps, the scientist can establish their credibility and build trust with business leaders, which is really the only way they'll succeed in selling their solar panels.

Now that we have gone through an example, let's break down several methods scientists can build and maintain credibility.

ESTABLISHING YOURSELF AS AN INDUSTRY AND THOUGHT LEADER

One of the best ways to establish credibility is to become an industry leader and thought leader. This is done by consistently applying for speaking opportunities at conferences, engaging on social media by posting your thought leadership, connecting with others in the community, and attending industry conferences. The type of content you develop and the conferences you attend will depend on your industry and your career goals. Countless times having my Ph.D. in the field, having high-quality work, and being a visionary in a specific sector have all been tremendous advantages in selling to customers. Being a keynote speaker at an industry conference not only generated sales for us but was also of high value when casually referencing my talk to new potential customers. Often, by presenting at these conferences, we broadly established our credibility in a context that was also approved of by many of our peers and customers' peers.

This act of establishing yourself as a thought leader is not only highly valuable at conferences but also in the context of more private meetings with any size audience or setting. Quick ways to establish yourself in smaller, intimate settings can be referencing your talk in your slides, having a slide that shows you at the podium of conferences, referring to your industry blog, and countless other ways that point to your knowledge and sustained leadership in the field. Many scientists I know may be hesitant to present and frame themselves in this manner, or to specifically call out their Ph.D. in their introduction. But if we had not positioned ourselves in this manner, the advancement of our company and ideas would have stalled. Being a thought leader is a constantly moving target that is always changing. Customers and audiences want to listen to the visionary leader in the field. Be that visionary leader. Be the most knowledgeable person in the room.

There are two fundamental components of being seen as a thought leader that no scientist can skip:

1. Producing high-quality work to back up your vision

    This means any content or work created by you, such as publishing research papers, presenting at conferences, and participating in other activities that showcase your knowledge and skills—all of which need to be of very high quality.

2. Staying up to date on the latest developments

    This can involve reading the latest research, attending conferences and workshops, reading blogs, watching talks online, and engaging with others in the scientific community. By staying current and up-to-date, scientists can demonstrate their ongoing commitment to excellence and maintain their credibility in the eyes of potential customers or clients.

We will assume that all our readers are hard-working and high performers that excel and stay up to date. If you skip out on these fundamentals and are caught having poor quality work or not being up to date once, it can tarnish your credibility substantially and sometimes permanently. The good news is that you are already doing a lot to establish credibility simply by engaging with the knowledge we're offering in this book.

BUILDING STRONG RELATIONSHIPS AND YOUR NETWORK: WARM INTROS

The size of your network is fundamental to establishing your credibility as an entrepreneur, scientist, or business professional. The more people you know and engage with and who view you as credible and a leader, the easier it is to convince others to trust you and do business with you. This is where warm introductions and referrals come into play. When someone you know and trust refers you to another person, they are essentially vouching for your credibility and making it easier for you to establish trust and rapport with the new person. This is true of collaborations, customer introductions, venture capital introductions, and pretty much any other type of introduction. A warm intro will give you a much higher chance of success in selling your ideas and products, so you should always try to make this your goal. Additionally, a vital ingredient to having rapid growth and spread of your ideas is "the network effect." The more people who are hearing, discussing, and debating your product or idea, the more people who are exposed to it without any effort from you. This also builds further credibility.

Networking and building relationships are essential to growing and establishing credibility in relevant communities. This can be done through various channels such as LinkedIn, Meetup groups, alumni events, industry events, and collaborating on research or industry projects. By actively seeking opportunities to share your knowledge and expertise, you can build a strong network of peers who view you as a credible source.

Note: the size of your network is only one piece of the puzzle. The quality or depth of your relationships is just as important as the quantity. Building deeper relationships with your peers through follow-up calls, interacting with the people in your network, and posting relevant content can help you become a credible source in their eyes. By understanding their needs and wants and why they are involved in the field, you can connect on a deeper level and build bonds that extend beyond professionalism. Building deeper connections requires consistent effort over a long period but is always worth it. Be open to the randomness of the world without knowing exactly where the connection will lead. Countless times, the person I thought would have the least value toward my goal has ended up being the most important customer, investor, or advisor in my career.

One example of when credibility in your network is at its strongest is when you have one of your industry-leading customers as a reference for a new customer. There is no stronger reference than that from one customer to another. By building this level of credibility and trust, you increase the chances of closing a sale. While networking and building relationships may seem like a lot of work, it's essential to maintaining credibility long-term and growing your business.

USING HONESTY AND HUMILITY

Selling as a scientist requires a unique approach that combines technical expertise with effective communication and relationship building. It also requires a unique balance of honesty, humility, and confidence without becoming arrogant or stretching the truth. Honesty will be required to build foundational relationships with your peers, audience, and customers. Honesty is crucial for building trust and credibility with your audience, and any lack of transparency can damage relationships and harm your reputation. There are times when being fully transparent and honest can put your product or idea at risk, but you must be creative to never overtly misrepresent or find a unique way to position to maintain integrity. If you are caught not being honest one time, it can be enough to destroy a relationship forever. Be careful if you feel you have to overpromise or commit to a feature release that is not fully ready to close a sale or investor.

Humility is a big part of the sales process for scientists. Being humble in your approach can help establish a connection with your audience and build credibility. This can involve sharing personal experiences or challenges, such as failures or struggles, so you can connect on a human level and show vulnerability. By being open, you can show that you are relatable and approachable, making it easier for your audience to accept and trust your ideas and expertise. Humility opens the door for them to accept your greatness. When I started this book, I told you a lot about myself and started with my failures, my fears about the big ORS talk, my struggles, and my concerns. This was meant to help connect on a human level and to show humility. If I started with "I'm the best," and "Only I can do this," you would not have connected in the same way to my messages in this book. Showing humility allows your audience to open up and receive your message because they are now on your side. Whenever you're selling, feel free to share some personal notes on challenges or difficult experiences as they demonstrate humility.

Confidence is also essential for scientists who sell. But you must find the right balance between confidence and humility. Being too humble can make you appear uncertain or unprepared while being too confident can make you appear arrogant or dismissive of others' ideas. Strike a balance between being open and honest about your challenges and weaknesses while effectively communicating your expertise and the value of your ideas. Many famous tech CEOs are seen as both confident and arrogant, which can help them grow a big company. However, this limits their ability in the sales environment to convince and coerce customers to pay. While it is possible to still be successful and be overtly arrogant, the best is a balance between humility and confidence.

FIGURE 7.1

Success Matrix for Confidence and Humility
subtitle goes here

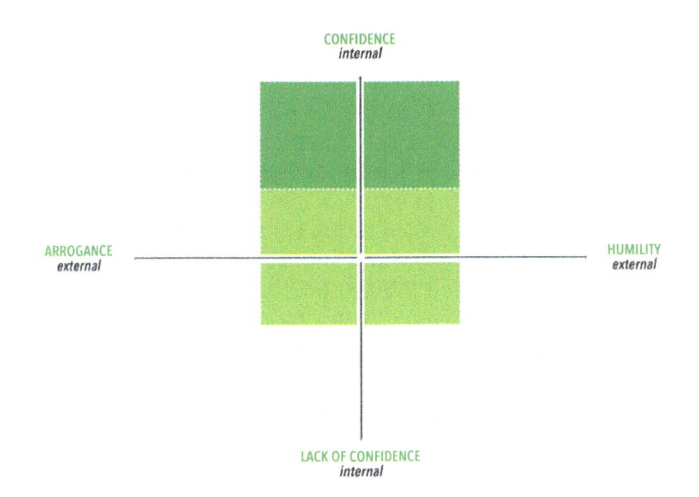

Figure 7. Success Matrix for Confidence and Humility

One last small trick in demonstrating humility is a common technique where one will try to show transferred credibility of your product, platform, or experience. For example, when I was selling enterprise AI software, frequently on our first sales call, we wanted to make it clear we were experts and that other leading companies were using us. But we never wanted to directly state this. So, we would plan several questions as part of the first call. One such question would be, "We are seeing that other leading Fortune 500 customers like 'Name-of-Company' are experiencing these types of issues; is this the same for you?" If it works for sales of $1M+ deals, it will work for you as well. Demonstrate your credibility but do it in a non-obvious way.

HAVE A LONG LIST OF RELEVANT AND RELATABLE STORIES

Audiences and customers love stories, which are the most effective way to inspire and convince others of your valuable ideas. Stories stay with people. What better way to leave a lasting impact on a customer than a memorable story demonstrating how you are agile or mission-driven? As such, make sure you have some pre-packaged stories to demonstrate your credibility in any sales situation. Imagine being in a conversation with a customer or an investor and a specific topic comes up regarding your technology, idea, or experiences, and you have the perfect story ready to go to satisfy their needs. This gives the appearance that you have a tremendous amount of experience, knowledge, and credibility. Believe it or not, it's okay to rehearse these stories. It is essential that when these teachable moments ripe for good storytelling occur in your life, you write them down. These stories you save and share are often the ones that end up shaping you, helping you grow, and influencing your judgment for your next important decision. I've started a habit of texting either myself or a close friend every time a profound story or idea comes up in my life. By texting yourself, you create a log that you can easily refer to at any time. Beyond a list of stories, another great practice is to log all your activities and achievements.

NETWORKING

Establishing and maintaining professional relationships can help you build credibility, stay current in your field, and expand your opportunities for growth. The following exercises are designed to help scientists improve their networking skills and build a strong network of professional contacts. Whether you are looking to attend networking events, connect with new professionals on LinkedIn, share your expertise, collaborate on projects, or follow up with contacts, these exercises will guide you through the process and help you achieve your networking goals.

1. **Networking Challenge** - Set a goal to attend at least one networking event or professional gathering per month. Make a list of upcoming events in your industry and commit to attending at least one of them. Prepare an elevator pitch that highlights your expertise and what you can bring to the table. Practice this pitch before the event so you feel confident and comfortable when you meet new people.

2. **LinkedIn Connections** - Set a goal to connect with at least five new professionals on LinkedIn each week. Look for people in your industry who have similar interests or expertise and reach out to them with a personalized message. Once you've connected, make sure to follow up with a message or a call to build deeper relationships.

3. **Expertise Sharing** - Look for opportunities to share your expertise with others. This can be through writing articles, giving presentations, or participating in industry panel discussions. Not only will this help you build your reputation as an expert in your field, but it will also allow you to connect with other professionals in your industry.

4. **Collaboration** - Look for opportunities to collaborate with other professionals in your field. This can be through research projects, joint ventures, or mentoring programs. Collaborating with others will help you build deeper relationships and provide you with opportunities to learn from others and grow your expertise.

5. **Follow up** - Follow up with people you meet at networking events or connect with on LinkedIn. Send a note or an email to thank them for their time and to offer any assistance or information they might need. This will help you stay top of mind and build deeper relationships with the people you meet.

SELF-REFLECTION EXERCISES:

Story Logging

If you are reading this book, then you are likely an overachiever and are looking to stand out from the crowd of other scientists, researchers, and engineers in your field. And there are likely dozens of activities you have participated in that demonstrate credibility, but you may not have written them down. It is easy to forget the smaller achievements that may not seem like they build credibility, but they do. It can be as small as being on a podcast on a particular topic, a panel, advising someone on a direction to take, mentoring a student, being a judge, attending a conference, writing a blog, doing a class project, or anything in between. No matter your career or personal goal, building credibility is essential to having success. And having a list of activities, accomplishments, or other relevant matters can instantly show your passion and fortitude. Is there a common thread you wish to develop a story around that can help you gain the trust of others at work? Are there common questions customers ask about your product? Write down one of those stories now and start the log! Stories will go a lot further on presentations and influencing as our brains are trained to remember stories rather than a linear set of facts.

Create A List of At List Five Extracurricular Items

Take some time right now and think about things you may have overlooked. Was it a lab presentation or a poster session for your school? Was there an event you attended related to one of your passions that can easily be connected to your story? Write down a list of at least FIVE extracurricular items not on your LinkedIn, resumé, or CV to start building your credibility.

COMMON MISTAKES

There are several common mistakes that people make when it comes to building and maintaining credibility in a sales situation. Some of these mistakes include:

- **Failing to establish expertise** - This can involve not staying up-to-date on the latest developments, not presenting high-quality work, or not effectively communicating one's knowledge and skills.

- **Lacking transparency and honesty** - This can involve not disclosing potential conflicts of interest, not being upfront about limitations or risks, or making false or misleading claims.

- **Not building and maintaining strong professional relationships** - This can involve not networking with others in the industry, not collaborating on projects, or not participating in professional organizations or associations.

- **Failing to clearly communicate value** - This can involve not presenting data on cost savings or benefits, not highlighting unique features or benefits, or not effectively positioning one's products or ideas in the market.

- **Too Many Words of Caution -** Be careful to not put in "disclaimers" everywhere when discussing your ideas and work. This is one of engineers' and scientists' biggest weaknesses. Detailed disclaimers actually break credibility for most executives and decision-makers. The question for the technical hammer: "If I give you the 10 million dollars you've requested, can you build a proof of concept that gets me to where I want to go?" The answer is, "Yes, absolutely," and then turn to their technical lead and be clear and direct on the next few points. Of course, their job is to push back and clarify but be clear, brief, and precise.

By avoiding these common mistakes, you can increase your credibility and improve your chances of success in a sales situation.

PUTTING IT ALL TOGETHER

Credibility is a crucial element for scientists to be able to sell their innovative ideas or products. It helps establish trust with potential clients and customers, differentiate from competitors, and increase the perceived value of your offerings. Building and maintaining credibility can be achieved by consistently producing high-quality work, fostering strong professional relationships, being transparent and honest, and staying current in your field. Studies have shown that credibility can lead to higher perceived value and increased trust in a product or idea, making it essential for scientists to prioritize building and maintaining their credibility.

CHAPTER 4

FRAMING

STORYTIME

*"From Product Failure to
Success With Framing"*

These were the types of comments we were seeing after a recent product launch. Unhappy customers, reliability issues, and loss of data. All really scary potential signs for our product line and our continued business. What were we doing to do?!

Fortunately, we remained calm and quickly got together with our leadership and advisors to find a way out of the scary and dire situation.

This was a big and unexpected product mechanical failure. How did this happen? We did rigorous testing, but somehow the customer was using it in an unexpected way that led to this failure. We had already paid for manufacturing for thousands of these products. This could be the end of our company...or so we thought. That was until we went back to the drawing board and FRAMED the situation in a new light.

Our recent results had shown us that we could get a majority of the product value from the one component collecting data rather than both. What if we didn't need both components and could just forget about the one that had been failing? This seemed to be a compromise that could actually work out for the best. Less product cost, eliminate a point of unexpected product failure, and we could still provide the core value to our customer.

However, there were different optics of how our customer would view this. That would prove to be difficult. If not handled right, this would obviously seem like a huge product failure, a risk, and a good reason to never work with us again. And we could not let that happen. So, we FRAMED the situation to our customer. We took the product failure and told our customer the following:

- This was actually an exciting and timely opportunity as we have advanced our machine learning and analytics capabilities. We were *already* planning to remove that additional component, and this was accelerating our roadmap.
- That means the same core value but fewer parts to manage, no risk of product failure anymore (at that point), improved user experience with fewer parts to manage—and best of all, we would replace every component for free, with no additional cost.

If we did not reframe this, it could have been viewed as a horrible product failure—the end of a business relationship. Except we took what could have been one of the worst things to happen to our company, and turned it into an exciting opportunity for our customer, reduced our BOM cost, and eliminated a major product failure. Framing saved the day yet again.

WHY YOU SHOULD CARE AND WHAT IT UNLOCKS

1. Framing can impact how information is perceived: By framing information in a certain way, you can influence how it is perceived and understood by others. This can be particularly useful in a sales or marketing context where

you may be trying to persuade someone to take a particular action, make a purchase, or believe in your ideas.
2. Framing can help you overcome objections and concerns: By framing information in a way that addresses potential objections or concerns, you can help overcome any obstacles that may be preventing them from making a decision.
3. Framing can help you create a sense of urgency or scarcity. By framing a product or service as being in short supply or as a rare opportunity, you can create a sense of urgency and motivate others to take action. This can be particularly useful in a sales context where you may be trying to encourage someone to make a purchase before it is too late.

Overall, framing is a valuable tool that can have a significant impact on how information is perceived and understood and can help you persuade others while building trust and credibility. If you want to increase the sales and growth of your idea, product, or company, make sure you are framing. A scientist who is an expert in framing can turn lemons into lemonade, even if there are no lemons.

PERSPECTIVES:

A Scientist's Perspective

As scientists, framing is another concept we rarely consider. Many scientists think along the lines of, "There is nothing to frame because the facts are exactly what they are, and that is exactly what will be presented." But in the real world and the business world, everything we see, consume, and buy is being framed in front of us. A product to one person is positioned and thought of completely different from the next person. In the real world, for scientists and engineers to stand out, we often cannot trust our instincts because we frequently present ourselves and our ideas as facts with no flexibility in the matter. Instead, we need to make ourselves a little (or a lot) uncomfortable, so we and our ideas get more visibility and have a larger impact. We need to frame our ideas, products, and presentations to sell more and faster.

A Salesperson's Perspective

From a sales perspective, framing is all about understanding the perspective of the person or people you're trying to sell to. You must first reverse-engineer what they know, what they want, who they know and respect, and who they would like to be. Framing helps you close the sale that you once couldn't. Framing lets you in the door you couldn't get in before. Framing allows you to be agile and sell your ideas to many. Suppose you have a mass-produced vase you want to sell. However, every person you sell to may be looking for a different color or style of vase. Before you sell to each person, you are able to customize the color and design for each buyer so it looks like the vase meets their needs. Framing is exactly this process. It lets you take an idea and customize it to the needs and wants of your potential buyers.

STATS THAT MATTER

- According to research by the Harvard Business Review, framing a product or service in a way that aligns with a customer's values and goals can increase the chances of closing a deal by up to 34%.
- A study by the Sales Management Association found that salespeople who effectively use framing and context in their pitches are 63% more likely to close a deal than those who don't.
- Another study by the Journal of Marketing found that framing a product or service in a positive context can lead to a 20% increase in sales.

QUICK WINS (M)

Fundamentals of Framing are SOSS:

- Stretching the facts and circumstance
- Omitting non-relevant details that can distract or take away from message
- Spinning the situation and reality so your idea and message are on top

- Suggesting or implying greater impact and relevance than may be currently accurate
- Using Analogies Can Be Extremely Effective

A really fast and simple way to get your audience or customer to understand and adopt your concepts can be analogies to something they are comfortable with.

DIGGING DEEPER

In selling, framing is critical to building the foundation of your value to your prospect and for your prospect to understand how to relate your knowledge and value to their knowledge and belief system. Framing is a powerful tool that can have a significant impact on how a sales situation is perceived and ultimately resolved. To better understand the value of framing in a sales situation, it is first necessary to understand what framing is and how it works. Framing refers to the way in which information is presented or communicated, and it can have a powerful influence on how that information is understood and interpreted by others. For example, the same piece of information may be perceived differently depending on how it is framed, whether it is presented as a problem or an opportunity, or whether it is presented in a positive or negative light.

In a sales situation, framing can be used in many ways to improve the outcome of a sales interaction. Here are just a few examples of how framing can be useful:

- Emphasizing the benefits of a product or service to the customers' unique problems. By framing the product in this way, the scientist can help potential customers see the value in the product and motivate them to make a purchase, so their concern around return on investment is clearly resolved.
- Overcoming objections or concerns a potential customer might have about a product or service, such as high initial effort required to set up and learn. By reframing their concerns to be small in the context of the much larger and long-term value, the scientist can help encourage them to make a purchase.

- Creating a sense of urgency or scarcity around a product or service. For example, if a scientist is selling a limited-edition product or a product in high demand, they might frame the product as being in short supply or as a rare opportunity the customer should not miss. This can help create a sense of urgency and encourage the customer to make a purchase "before it is too late." This topic is so important, there is an entire chapter dedicated to it later in the book.

- Highlighting the competitive advantage that the product or service can provide, if we are aware that there are many other popular solutions to the problem. By framing the product or service as a unique solution with specific advantages over other products, it can help a business stand out from the competition.

In the real world, for scientists and engineers to stand out, we often cannot trust our instincts as we present ourselves and our ideas. Instead, we need to make ourselves a little (or a lot) uncomfortable, so we and our ideas get more visibility and have a larger impact. We've boiled down a simple framework that will help you promote yourself, your ideas, and your work in a more exciting and engaging way. The four components of Framing are: Stretch, Omit, Spin, and Suggest (SOSS).

STRETCHING

The first lesson of SOSS is **Stretching**. Stretch yourself, your work, and your accomplishments. Stretching is the act of increasing, growing, or embellishing the facts around you or your product that can help you to better influence or sell your idea. Now I am in NO MEANS suggesting you lie, but hyping up your work with marketing "power words", or focusing on certain details or statistics over others, can be the difference between landing the sale...and *not*. There is no harm in positioning yourself and your efforts in the best light possible. As scientists, it can be in our nature to state our case dryly and clearly, not wanting to be loud or selling or overselling our case. But stretching our ideas and value is what we must do to be successful in selling. And more importantly, others you are applying against, trying to outcompete, are already doing it! The sooner you start to stretch, the sooner you can compete.

The one area where scientists often fail to stretch is in describing themselves and their accomplishments. Here are a few examples of different situations where scientists may fail to stretch from real world examples of scientists putting together their resumés and applying for jobs:

- If we are reviewing your resumé, and you just finished your Ph.D., that is at least five years of experience in your field right there. Many Ph.Ds. neglect to include their research experiences or their Ph.D. as time accrued in the field.
- If you had an honorable mention but are not sure if it is listed online, then and include it rather than worrying about whether or not it can be verified.
- If you were head Teaching Assistant and got a scholarship, that means you wrote and received grants and helped to develop a course.

Unfortunately, your research, credentials, experience, or personal accomplishments will often be slightly less (or drastically less) than what is requested from opportunities. This is okay because no candidate meets all the seemingly endless qualifications desired. Companies struggle to find the right people and the right talent. With most companies, you'll need a certain amount of training and on-the-job learning, anyhow. They are looking for a fit with the team who also happens to be smart, talented, and hardworking. You'll need to be proficient in some of the skill sets required, while willing to learn the rest. So, the first lesson here is that you should reach for jobs and opportunities beyond your strictly defined qualifications.

Another area where scientists often fail to stretch is in describing their status, product, and capabilities to customers or investors. Here are a few examples of this, along with accompanying solutions:

- BEFORE: We weren't able to hit the deadlines.

- AFTER: We are seeing amazing trends in the results. The increased customer growth has forced us to push back current deadlines.
- BEFORE: We did a couple of small pilots of our platform for several customers.
- AFTER: We have deployed our platform to over 5 customers and are on track for more than 25 by the end of the year.
- BEFORE: Our product has only 50 current users on the platform.
- AFTER: Over 250 wearables have been delivered to customers, and we are on track to reach 1000 by the end of the quarter.

Generally, to stretch your traction, ideas, and value, the following "pro tips" can be used to help get you started:

- Stretch by projecting where you will be, even though there will be some guesswork involved.
- Stretch by including additional accomplishments and positives that either help to alleviate or distract from any challenges.
- Stretch by using a large, vague number or by using the words above or over, such as over 1000 units shipped, even if it is only 1001.
- Stretch by assuming early commitments or questionable shipments are fully committed, such as sales or deployments. If that is your expectation or goal, then you can include it.
- Stretch by using numbers or metrics that appear largest. If you have only a few customers or publications, maybe look at the number of products deployed, or daily uses, or citations to seem more grand.

OMIT (UNNECESSARY DETAILS)

When you are selling yourself, your idea, or your product, I know you want to be honest and present the "entire truth", but nobody knows what that is except you. Often the whole truth is distracting and not relevant or important to your audience. When it comes to that difficult question, only mention those positive items, and feel free to omit a couple of the mistakes or troubles you may have had along the way. Now this will be no easy task to leave out details when communicating or giving a presentation for a scientist or PhD, especially since those details are SO important, and there are SO many of them, and everyone will need to know every single one of them. This is part of our training, and essential to be a successful researcher, when every experiment has to be done meticulously, and every word of the methods perfectly crafted, and every detail of the results dotted out. This is especially common for scientists or technologists turned entrepreneurs, as we very often will fall in love with the solution and the technology we have built. And so when given a chance to talk about it we focus here, when in reality the investor or customer a favor because all they really care about is the value that you, your idea, or your product creates and that you understand their problem.

I can recall numerous group meetings where the smartest minds in the world would become obsessed with what color the dot for X experimental group should be or the line thickness of Y plot axis. And suddenly you begin to see a waste of minds arguing over unimportant minutiae. I also experienced this early in my entrepreneurial career. When building a company, creating a pitch deck is one of the first you create so you can present to customers and/or investors what you do. As a scientist, meticulously dedicated to details I would make sure to include every feature and detail of our product. And these pitches failed miserably. When you offer a sea of details it can be tough for the customer or investor to focus. Imagine a photographer trying to focus on a specific leaf, in a tree. No matter how much you try, those details will be lost and the ability to focus is gone. This is in practice what we need to do when communicating and storytelling as well.

When you are omitting, the recipe can seem quite simple, though, in practice, it can be quite difficult. Your goal is to only include the essential facts that communicate your story clearly. This means you're focusing on only one message and one linear story, and you're only including details that support that story. Our expertise on the topic we are presenting will often make omitting difficult. As a result, we may be too close and fully immersed to see outside of the details and what is non-essential. You may be able to put together all the details in your mind, but it's likely your non-expert audience will only be able to capture the main essence of your message.

Omitting is also important when those unnecessary details take away from your story or discredit your idea. I often ran into this while presenting our company to venture capitalists. I would need to concisely describe a pilot program for our technology, and rather than just focusing on the one or two impactful outcomes relevant to the story, I would share other anecdotes and details that pulled focus from my main point.

Building on this, omitting also applies to answering questions; we only need to answer questions in a way that highlights the value of our product/service or shows the depth of our expertise. A famous quote from Henry Kissinger exemplifies this best: "Does anyone have questions for my answers?" When omitting unhelpful details, we can maximize the impact of our message.

SPIN

Selling with spin is the art of repositioning facts and statements to better align with a customer's or audience's perspective. Any fact or statement can be repositioned; it simply requires some effort and is a bit of an art form. For example, a seemingly negative result can be spun into a positive if presented in a different light. A terrible product failure can be turned into "an amazing learning opportunity." In the opening story of the chapter, the product failure was spun into a positive and a benefit for the customer. By reframing failure as a way to improve future products and better serve customers, the negative becomes a positive.

Another way to spin a situation is by turning a delay in the delivery timeline into something positive. Instead of focusing on the delay, you can frame it as high demand from customers. This approach reframes the delay as a good thing, highlighting the high demand for the product or service and how it's positively impacting the company.

Scientists often need to learn how to spin their reality to be successful in sales and convincing others of their ideas.

By understanding how to spin facts and statements, scientists can better communicate their ideas and research to potential customers, partners, or investors. By using spin, scientists can present their ideas in a more positive and compelling way, making it more likely that others will be convinced and interested in what they have to offer.

SUGGEST

The last rule of SOSS is **Suggest**, which involves a bit of nuance and subtlety and is more of an art than a science. We're not dumping all our information on the listener all at once; we need to lead them along and take them on a journey. Create a sense of suspense and intrigue. As we present or sell, we want to start by offering images and concepts that hint at important ideas, credibility, or context, without actually saying them. Like a magician guiding your eyes away from the trick, you must elegantly guide the audience and the conversation to where you want it to go. The better you get at this, the more it will be seamless; your audience will never know.

For instance, when we were presenting our sales deck to customers, we needed to show credibility that we had actual users on our platform who were having success. This was difficult in the beginning when we had very few users. To counterbalance this, we used wording that suggested we had many more than the few users we showed. In our deck, we added a "Select Customers & Success" slide. When we presented this, we always made it a point to say we were highlighting select customers. These "select customers" led to the suggestion that we had many more.

Let's say your new company has taken on more customers than it can handle. When it's time to offer a delivery timeline, rather than saying your company is behind on production, you can say your company's product has been sold out for weeks, and you can only ship in 6-8 weeks. The next time you are presenting to a customer, a boss, or an investor, think about what messages you may want to suggest and how you can subtly include them in the presentation.

SOSS CLOSING

You will need to be cautious as you navigate the delicate line between truth and fiction. I am in no way saying that we should lie, nor should you present something wildly off-base from reality. But scientists like me often tie themselves to delivering every detail in the name of presenting "truth." As if providing every detail helps us maintain our moral code. While this seems honorable, it's not helpful to the customer. You don't want to exaggerate or skew your data, but you can still tell a barebones (truthful) story of your results that shows you understand the customer's problem and that your product or idea has value in solving their problem.

In April 2023, SpaceX's new Starship launched and exploded in the sky 4 minutes after liftoff. SpaceX didn't go into the minutiae; they simply told the media and its partners that the launch was a success because they learned a lot for next time. Talk about expert spin and omitting unnecessary details!

Scientists should learn from SpaceX and feel free to do the same when appropriate. By considering the SOSS framework, you are guaranteed to improve your messaging to external audiences. In reality, nobody cares about you, your invention, or your hard work. They care about themselves and the potential benefits your invention can provide them.

However, you must take into account that everyone is different and unique, meaning we must always use framing for best results.

FIGURE 8.1
SOSS Framing Overview
subtitle goes here

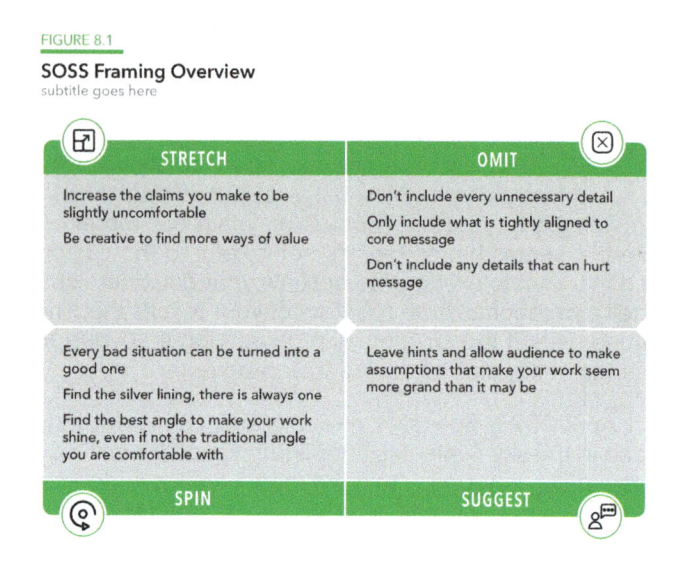

STRETCH	OMIT
Increase the claims you make to be slightly uncomfortable	Don't include every unnecessary detail
Be creative to find more ways of value	Only include what is tightly aligned to core message
	Don't include any details that can hurt message
Every bad situation can be turned into a good one	Leave hints and allow audience to make assumptions that make your work seem more grand than it may be
Find the silver lining, there is always one	
Find the best angle to make your work shine, even if not the traditional angle you are comfortable with	
SPIN	SUGGEST

Figure 8. SOSS Framing Overview

USING ANALOGY TO REFRAME FOR FAST RESULTS

Think back to the last memorable speech or TED talk you watched. At some point, while talking about a new, creative, or innovative idea, the speaker used an analogy to make a complex topic seem simpler. Perhaps they did so by comparing their novel idea to something more common from another field. As humans, we are constantly doing these kinds of comparisons consciously and subconsciously throughout our day because it helps us better understand the world around us. This form of cognitive processing is called analogical reasoning. Numerous studies demonstrate how analogies assist in explaining complex scientific topics to a broader audience. Analogies remain one of the most powerful tools for memory recall, as it forces complex mental patterns and visual imagery.

Analogies like metaphors and similes are also how new innovations and many of our favorite creations come to fruition. Some of the biggest ideas or innovations are simply applying a technique or methodology from one field. Ask any venture capitalist as well, and they would be more likely to invest or believe in the success of a company if you are applying something that already exists but into a different area. A venture capitalist is much likelier to grasp and remember your concept and perhaps even invest if you use elevated language like metaphors and similes (e.g., "We *are* Uber for Veterinarians," a metaphor or "We are *like* Airbnb for Houseboats," a simile).

I have used analogies in my career to much success. For my research during my Ph.D., I looked at two complex and different fields, the gut microbiome and orthopedic disease. By immersing myself in both these worlds, I was also able to see how techniques and tools applied to research in one field, would be able to provide unique innovation and insights when applied to the other. For example, I was able to apply some of the machine learning concepts used in gut microbiome research to my data in the bone biomechanics world to find key results in my data.

In our startup, we struggled to find an appropriate analogy that would not only resonate but appropriately simplify and explain. When we were fundraising, we would often see startups pitching themselves as "Uber for Planes" or "Netflix for Edu-videos," and often the pitch worked. These companies were able to capture the imagination of venture capitalists while becoming ingrained in their memories. By analogizing, these startups were also associating themselves with a successful company with a proven track record.

Analogies also work really well for speeches and for formulating jokes. Comparing two things not normally associated with one another can be comedic in nature (e.g., "He was *like* the Michael Jordan of Accounting Math").

When I described my research on the gut microbiome, an incredibly new topic, by telling my audiences to first, "Imagine if for millennia we were unaware of another organ inside our body, dictating how we behave, how we digest and get nutrients, and we suddenly discovered it was there the whole time." That proved to be a powerful analogy that stuck! If I asked an audience member to briefly describe the gut microbiome to a stranger, they would offer something along the lines of, "It is essentially a whole new organ that has a dramatic impact on everything."

When creating a new analogy try to make it visual and incredibly simple so it resonates with as wide of an audience as possible. There are two types of common analogies used:

1. **Analogies that identify identical relationships**. "Black is to white as on is to off." In this example, they are both opposites.

2. **Analogies that identify shared abstraction**. "Raising children is like gardening—nurture them and be patient." Essentially, we are comparing two things that are technically unrelated to draw comparisons between an attribute or pattern they share.

SELF-REFLECTION EXERCISES

Practice Your Own SOSS

Think of a recent experience where you had a problem or even a minor success. As you start to recall this experience, think about it as if you would explain it to your boss or peer. Write out the story in one brief paragraph, and make sure to include any necessary details. Afterward, go through and highlight the following:

1. Any details in the story regarding results or stakes or timeline, where you could expand the truth a little (Stretch). It can be that instead of a month you were able to do it in a fraction of the time previously, or instead of increasing revenue by $100k, you were able to nearly double it.

2. Any unnecessary details that can be omitted because they do not strengthen your story. Remember you are trying to communicate one simple, clear message.

3. If it is a negative or problem in the story, is there a way we can spin to improve the messaging. If there is a positive can we spin it to be even more impactful.

4. Any areas where you can suggest something extra or added value to your audience.

After you have highlighted areas for SOSS, rewrite your one-paragraph story. Review the two versions and see which one you find more memorable and well-written.

Analogies Please

Think of a complicated scientific topic that you understand well. It can be related to your life, your work, or a passionate hobby. Imagine you need to explain this topic to someone completely unfamiliar with it. What do you do!?

First, we need to simplify the topic at its essence. What details can be omitted? What is the central message you want the listener to grasp?

Next, as we begin to craft what we're teaching, we need to insert elevated language like analogies, metaphors (e.g., "time *is* wasting"), and similes (e.g., "our product is *like* the Uber of speedboats.) that capture the essence of your topic. Elevated language helps the listener make connections in their brain between something new and unfamiliar (e.g., this topic or even your product or service) and something familiar (e.g., a familiar sports story about their favorite basketball player or recognizable metaphor, "time is money.").

Once you have your beautifully crafted analogy, it is time to test. Tell the original concept to a friend without using the analogy. Then tell the same concept to a different friend using the analogy. See if either or both understand the concept. After a week, go back to your two friends and see if they can explain the concept. You may be impressed by how long an analogy can be locked into the memory.

Finding the Good in the Bad

Think of a recent failure that you were really upset about. I know it seems terrible, but in nearly every difficult situation, we can find a positive or a lesson learned. That is the simple goal of this exercise. Write down three ways this failure or problem can be viewed as a positive. Try this again with another recent problem. After doing this a few times, you will start to build muscle memory and confidence that you can always find a way to spin something positive out of something difficult or painful.

COMMON MISTAKES

1. **Failing to consider the audience**: Ten people may respond ten different ways to the same message depending on their beliefs, values, and experiences. Failing to consider the audience can lead to ineffective or even counterproductive framing.

2. **Using biased or misleading framing**: Always be honest and accurate when framing information, and avoid using biased or misleading framing that could mislead or deceive. Doing so can damage trust and credibility and can ultimately backfire.

3. **Being too subtle or too heavy-handed**: Find a balance when framing information. Being too subtle can result in the message being missed, while being too heavy-handed can be off-putting or even manipulative.

4. **Failing to consider the context:** The context in which information is presented can significantly impact how it is perceived and understood. Failing to consider the context can lead to inappropriate or ineffective framing.

5. **Stretching too Far or Omitting too Much**: This requires a delicate balance and is a bit of an art. If you are new to this, it can be easy to go too far and misrepresent the truth and damage trust and credibility

PUTTING IT ALL TOGETHER

Framing is a crucial element in the selling process. It helps establish the value of what you are offering to your prospect. By framing yourself, your accomplishments, and your value in a specific way, you can more effectively communicate the unique benefits of what you have to offer and convince others to buy your product or idea.

CHAPTER

5

CREATING MOMENTUM & URGENCY

STORYTIME

"Fundraising With No Momentum and No Urgency"

Fundraising was a constant struggle while working on my startup. And as soon as we closed a round of funding, it seemed we were already behind on the next one. This constant push and pull meant it felt like we never had momentum or any urgency. When we were raising our funds, we were often honest about our positioning, "Yes we are just starting our round. We don't have any commitments yet but hope to." The response from investors was often, "We would like to stay in touch as things progress" or "Let us know when you are going to close," both of which are nice ways of saying "no." We didn't solve this problem until we realized the power of creating momentum for our story and a sense of urgency for investors to act now. One way we did this was showing that our sales and funding started in a certain place but with appropriate funding would grow ten times that amount. We went from zero commitments to a 50% conversion rate. What created a sense of urgency by telling investors we had a limited amount of funding for this round that was accelerating. They needed to decide immediately if they wanted to join our rocket ship. Armed with momentum and urgency, we were able to close millions of dollars in funding.

WHY YOU SHOULD CARE & WHAT IT UNLOCKS

Momentum and Urgency

Momentum helps people believe in your idea, provides credibility, and inspires confidence your idea will be successful. Urgency is what forces them to decide or act now.

Think about the last time you were asked to make a decision or choice. What about the last time you saw a deal in a weekly ad. Think about buying a car and the salesman telling you their special sale ends soon. Think about having to decide on a job offer or which Ph.D. program to accept by a certain deadline. In each of these situations, a timeline forced you to make a decision and created a sense of urgency to act. This urgency is what we as scientists need to learn to harness to urge others to make a decision in our favor.

If you can practice and harness momentum and urgency, expect more people to believe in your ideas and to take action.

PERSPECTIVES:

A Scientist's Perspective

As scientists, we probably haven't considered momentum and urgency unless it involves making sure we finish sample analysis on time or making the deadline on a major grant. It can be difficult to realize the importance of the concepts when your research may take you 5 years to reach definitive results, and you think "Hey what is another hour or week". Momentum can feel like something we don't have and never had as scientists, especially when we are aimlessly working towards an unknowing end date to our Ph.D. and our research. Urgency is another topic scientists may not feel comfortable with, outside of their own head, as often times we may be passive and accept external influences. We accept urgency from deadlines or our advisors or managers, but we rarely use urgency to our advantage.

A Salesperson's perspective

We always use urgency to our advantage. In sales, momentum and urgency are the twin forces propelling us towards our goals. The moment one deal is clinched, we're already hot on the trail of the next. This unyielding cycle of momentum is our lifeblood, our rhythm, our norm.

Contrasting with a scientist's approach, our use of urgency isn't passive. We don't merely accept it as a product of external deadlines or circumstances. We actively manufacture it, wielding it as a powerful catalyst for decision-making.

Leveraging urgency and maintaining momentum is our craft. We're adept at spinning narratives that compel attention and action. We create a vivid picture of a journey that's just beginning, full of potential and ripe for growth. We don't invite complacency with phrases like "We're just starting our round." Instead, we spark interest, painting a picture of a unique opportunity that's gathering pace.

In sales, we're not just along for the ride. We're the drivers. We generate momentum, create urgency, and harness these forces to propel us forward. As salespeople, we thrive on this, turning potential into reality, one closed deal at a time. This is our advantage, and this is the power of sales.

STATS THAT MATTER

- A study by the Sales Executive Council found that sales reps who were able to create momentum in the sales process were 3.4x more likely to close the sale.

QUICK WINS

Creating Momentum to Gain Buy in

When referring to your work or success, never talk about it in a static state. Show where you were and where you are projected to go to create a convincing story and inspire success in your idea.

Use Urgency to Ignite Action

When given the choice to act or wait and see, most people will choose to wait. Create urgency in your pitch to encourage your audience to take action and make sure to highlight the consequences of inaction.

DIGGING DEEPER

Momentum

Whether you are on the couch eating Cheetos, sitting on a lounge chair on the beach sipping a mimosa, or anything in between, you are generating momentum...but maybe not the healthy kind. To quote Newtonian physics," an object in motion stays in motion." This also applies in sales and convincing others of your opinion. No company wants to be the first to purchase your technology. No conference wants to offer you your first-ever speaking event. The world wants to know that this is the exact right moment for you, and you have great momentum pushing you to the top and even further. For your customer they want to know your amazing track record and path that have landed you at their door. For your audience at your TED talk, they want to hear how all the prior failures and lessons have landed you to arrive at this groundbreaking innovation.

We need to leverage this principle of momentum in every situation where we're trying to convince someone of something.

Essentially, no action or situation happens in isolation. You are not just presenting your technology to this one potential customer. You've had numerous successes and opportunities that led you to this exact presentation at hand.

You are not just presenting your award-winning findings; there were trials and tribulations before, and there is an even bigger discovery on the horizon.

When presenting traction or facts, never present them in isolation of time. You must put your current situation in the context of what has happened prior and what will happen next. Once you have a before and after, make sure to show that things are growing and expected to show exponential growth after your most current update. Think of this graph as the timeline you want to show:

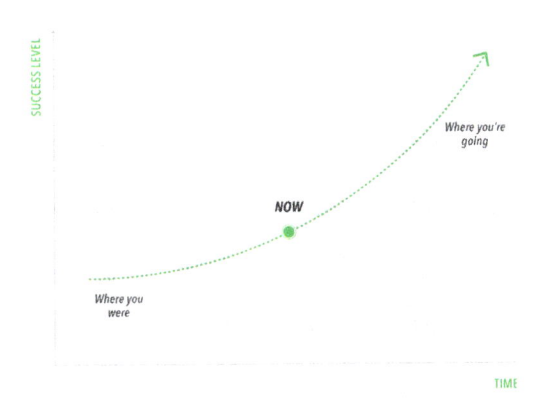

Figure 9. The Momentum You Already Have

Creating a momentum timeline in partnership with your customer can also be extremely helpful in working toward a future goal and alignment. Try setting milestones of 1 day, 7 days, 30 days, and 90 days to show the momentum journey you are planning together.

Lastly, momentum may help you to build credibility with your audience. When you discuss your ideas or showcase your work, you are asking your audience to take a leap of faith to believe in you and your concepts. Some people in the audience may inherently trust you and have faith in your idea. But most who are not aware of your work and ideas will be skeptical because trust has not been built yet. You may be able to bridge that gap with your audience by showcasing a story of where you started and where you are going.

Urgency of Now

Timing is everything in life and is probably the most important factor for success. From making the career move because you happened to see a sign on the road or bumping into someone at a networking event, to meeting your spouse at a bar. The concept of time, and specifically *why now* can be a powerful concept when used appropriately. In fact, the Greeks have two words to describe time.

- *Cronos* is time kept by our clocks
- *Kyros* is the concept that some moments are more important than others

When conducting a sale or trying to convince someone of your opinion, time can be used to accentuate and enhance your argument or message. While **Cronos** can be valuable and is necessary in establishing timelines for a sales deal or a presentation, **Kyros**—this moment is incredibly unique and needs to be taken advantage of—is what really drives people to make decisions.

In sales, time is often used to help force a decision and to raise the stakes or sense of urgency. In fact, in sales, if the urgency of now didn't exist, nobody would purchase anything. People inherently will wait to weigh any decision, especially customers. But if my problem has to be solved now or the deal is going to run out soon, then I will be more likely to purchase. Sales professionals know this and experts use it regularly to close deals fast. They provide some type of time urgency and why you need to take action now, even though the very next month there will be another deal.

Scientists and engineers can take this lesson to heart when trying to close a sale or when trying to be persuasive in a presentation. Here are a few tips and examples to apply in relevant situations:

- When trying to apply the urgency of now in your talk, the beginning and end are often ideal times to highlight the importance of the moment. Likely your research or your discovery are particularly important to the world right now or you would not be wasting your time. Make it memorable by focusing on the stakes of what the world would be like without your invention, and what happens if your invention isn't adopted. Paint this picture with vivid imagery and memorable statistics where appropriate. If we were to give a presentation on climate change, but don't apply the urgency of now, it would fall flat, and we'd wait until 2350 to take action.
- When trying to close a sale, there are numerous ways the urgency of now can be applied.
  - First, if there isn't a timeline applied to the deal or the process, then one needs to be established right away (*Cronos*). You can do so by simply asking what steps are needed to go from where you are now to a signed deal. Once a timeline is established, you will inevitably need to motivate them to decide now or soon (*Kyros*). Common techniques I've applied to help accelerate decisions include: a limited-time opportunity, the deal's expiration date, "another customer will take your available slot", "we've already allocated our inventory for next year...but I'll see if we can squeeze you in", "my boss is asking me to close this today" ... and so on.
  - Second, setting a deadline for the customer to make a decision. You can do this by highlighting the benefits of making a decision quickly, such as getting access to exclusive deals, discounts, or limited-time offers. By setting milestones, you can keep the customer focused and motivated to achieve their goals. This will help them stay on track and not get distracted by other things. Setting milestones also helps to keep you and the customer accountable.
  - Third, highlighting the potential consequences of not taking action. For example, if a customer is considering a new software, you can highlight the potential risks of not switching to the new software such as security vulnerability, missing out on new features,

or not being able to support its current business requirements. By setting clear milestones, both parties can stay focused on their goals and take steps to reach them in a timely and efficient manner.

SELF-REFLECTION EXERCISES

Creating Momentum in Your Life

Think back to your last few presentations where you were trying to convince another party that your opinion is the right one. Did you communicate the current momentum from your research, product, or idea? Did you present that you had already accomplished a lot, and that you were on the cusp of an exponential growth in achievement? If not, review what you were presenting on and rewrite it by adding in the concept of momentum. Highlight where you started and where you're going (momentum). Test out both the original version and the new version on your friends and colleagues and see which they prefer!

Exercise: Using Urgency in Your Life

Boy, it would sure be great if I could convince my friend to go to the movies with me today. I wish there was some way I could get them to understand the importance of going there now. Urgency is the recipe!

Think back to the last time you tried to convince a family member, friend, or significant other of your opinion or to join you in your preferred activity. Did they accept it or push back? What tactics did you deploy? Could you have gotten them to say yes or do it sooner if you tried to create urgency? The next time you find yourself trying to convince another party, use urgency and why this moment is special for what you are proposing.

COMMON MISTAKES

- Starting Flat: When creating momentum, you may believe, "I just started this idea or this job and have nothing to show." But that is wrong. Dig deep, and you will be able to find a story that generates momentum and growth—but you may have to be creative.

- Projecting Far and High: Often you may not stretch how far you will go with your idea because it does not (yet) feel plausible, and instead you shoot lower. To show where you are going you will have to expand your vision and expand the realm of possibilities. Dream big to show how far your idea will go, even if it makes you uncomfortable.
- Decision Date: Not making a timeline soon enough to create urgency. Timelines should be adaptable per the boundaries of each situation. It is best to force action when the timeline may make both parties a little uncomfortable.
- Timeline creep: It can often be uncomfortable to create a timeline and encourage a decision, which will sometimes cause individuals to request an extension. It is best to keep the timeline, or if it does change, make sure you are getting additional value due to the compromise.

EXAMPLE: BEFORE / AFTER

Check out the examples below and decide which sounds more appealing and exciting:

Before: We have sold 1000 units this year.

After: We beat our sales expectation, selling 1000 units this year; we expect continued month-over-month growth of 100%.

Before: We have demonstrated that our therapy has a 20% cure rate.

After: We have made several breakthroughs this year, doubling our cure rate to 20%, and we plan to double again by the end of next year.

Before: We would love to have you invest in our round if you choose to.

After: We are nearly 75% subscribed to our funding round, with many more expected to subscribe this weekend. If you would like to invest, this weekend will be the next chance to join before the round is closed; otherwise, you'll miss this groundbreaking opportunity.

I'm sure you will agree that the "After" statements convey a stronger message and are more exciting. Try it out for yourself!

PUTTING IT ALL TOGETHER

An object in motion stays in motion. We want you to be that object in motion to whoever you try to persuade. When we are selling ourselves, our idea, or our research, we are asking something from others. We want our audience to take action or make a decision today. By generating the urgency of now, we can authentically coerce our audience to take action now in our favor. The sooner you master the skills of creating momentum and urgency, the sooner you will be ahead of your competitors.

CHAPTER 6

PRESENTATIONS

STORYTIME

"One Presentation to Rule Them All"

We were in a factory presenting our technology to the plant manager and dozens of employees from all different walks of life, ages, backgrounds, education levels, and language proficiency levels. We had closed the sale a few months back, but this presentation was going to commence the deployment of our wearable technology. We had only one opportunity to engage the audience and communicate our value proposition clearly to excite managers and employees about the effort that would be required of them. The plant manager gave us a 30-minute window to speak to staff so factory lines wouldn't be off for too long (a huge expense).

However, in the weeks leading up to our presentation, we had been working hard but struggling to get the technology to work perfectly. Not much effort had been put into the on-site presentation. Our attitude was, "We have great technology. it will help solve their problem. They will *obviously* understand it and be excited about it."

Unfortunately, we miscalculated, and as a result, it led to an initial disaster. The managers and staff were not engaged from the start and failed to understand the purpose of why the technology was being deployed. The next month we saw a deterioration in the daily deployment of our technology as well as the proper usage. We had learned our lesson—we did not understand our audience or their perspective well enough. Our material was not put into crisp, concise messages for anyone to understand. And we did not explain or engage why the front-line workers should get excited about the technology. Sure, the CEO and CSO were excited, but that does not mean their enthusiasm would trickle down. In our next deployment, we went back to the drawing board, reworked our slides and messaging, and piloted the presentation virtually with staff to ensure it would hit home. As a result, our next deployment went incredibly well; workers were engaged and excited, management was happy, and our technology was able to scale within the facility and the company. The lesson: a presentation is never just a presentation. In this example, because our presentation went poorly, our technology was not being used or having an impact on our customer. They were losing business, which could have caused us to lose our startup. This same message should resonate for every presentation you give.

WHY YOU SHOULD CARE & WHAT IT UNLOCKS

Presentations are (still) the cornerstone of communicating ideas in a professional setting; you might as well get good at them, so you can excel your way up the corporate ladder and persuade others to buy your brand, products, and ideas, and give you money. It will allow your ideas to spread fast, and further build your name and reputation. Successful presentations are a cornerstone of a successful career, and success with all types of sales.

PERSPECTIVES:

A Scientist's Perspective

As a scientist, we probably haven't thought much about the importance of a presentation. Typically, it's a chore—something our manager may have asked us to put together, something our professor is showing us, or a last-minute scramble for an event or meeting. Our thoughts on presentations are likely binary, "My material is in presentation-format, or it isn't." A presentation is not simply taking your results and putting them on a slide. There is much more to it than that, and those that master this skill will excel further in all their endeavors. Scientists should think of presentations as a tool to convince and persuade others with their ideas.

A Salesperson's Perspective

As salespeople, we understand that a presentation isn't just a sequence of slides - it's a performance, a chance to captivate, inspire, and most importantly, to persuade. In our world, the best presentations don't necessarily get to the end. They don't need to because somewhere in the middle, we've sparked a connection, ignited a discussion, caught the twinkle in the audience's eye. We've transformed a one-way data dump into a highly interactive and engaging conversation. The best presentations, like great stories, inspire their audience to action. When you see that spark of inspiration, that moment when your audience connects with your idea, seize it. Don't worry about reaching the end of your slides. Your goal isn't to finish your presentation; it's to close the deal, to ignite that spark of understanding, enthusiasm, and excitement in your audience.

STATS THAT MATTER

- According to a survey by Forbes, 60% of customers say they are more likely to buy from a company that provides them with a clear and concise presentation.
- Several studies estimate that nearly 30 million PowerPoint presentations are given every day around the world. That's almost 11 billion per year.

QUICK WINS

Simplicity Is Always the Right Choice

Keep it simple—don't overload your audience with wordy slides and lots of complex images. Refine your presentation to the point where even a toddler can understand it. When you make your presentation and slides simple, your audience will feel smart and satisfied, making an even bigger impact.

Preparation Is Always Key to a Successful Talk

If a talk seems spontaneous and fun, then it's a good bet that the presenter prepared a great deal in advance. The allusion of spontaneity is a result of practice. The first part of preparation is fully understanding your audience and their unique perspectives and wants. Next, you want to make sure your presentation has a singular focus. You'll also want to include a great story with foundational elements and a call to action.

Start by outlining all these elements, creating the titles of each slide, and ensuring your presentation has a logical flow.

Engage Your Audience from the Start and Don't Stop

To get your audience to listen and digest your talk, you need to excite and engage them from the very start. Questions, interesting facts, and personal stories all help do this. Once you get them hooked, you need to continue to reel them in. Have planned and practiced ways to continue to engage.

DIGGING DEEPER

Make it so simple a toddler can understand it...literally

This strategy is why I was able to win the 3-minute thesis award competition at Cornell University. Now I may be exaggerating some regarding the level of a toddler, but essentially, a person with limited to no knowledge in your field, with limited relevant education, should be able to understand your talk. For me, I made my thesis dissertation so simple that my grandma—with no college degree, no background in science, and limited technical understanding—could fully understand it. That's right; my grandma had no problem understanding my summary on the pathways by which the gut microbiome can influence orthopedic disease.

This level of simplicity should be a goal in all your future talks because it will increase the likelihood that your talk will resonate with your audience, and guarantee your message is understood. Can your talk pass the Toddler Test™.

Whether it's your grandmother, your goofy cousin Tim, or a new colleague, when you practice your presentation, make sure you take note of the kinds of questions they ask. Can you talk it through with them and find a new way to explain it more clearly? A fresh outside perspective can be eye-opening. When you have spent thousands of hours on an idea, it can be difficult to decipher what is easy to understand and what isn't. The "Toddler Test" will remove any fancy lingo from your presentation and force you to simplify your language for a more general audience.

The point here is to present to someone inexperienced in your field. This allows you to boil down the essence of your talk and/or to find deep-rooted analogies that anyone can understand. I often see talks full of jargon and complicated details, leaving even highly-educated people lost in the wind. At most scientific conferences, I estimate that no less than half the audience is dreaming of cotton candy rainbows and the great outdoors.

You may get frustrated at the significant effort it takes to pass the Toddler Test, but the payoff will be high because most, if not all of the audience will understand your message. After all, what is the point of preparing a talk if no one understands what you're talking about?

Make Sure Your Audience Feels Smart

There is no greater feeling than going to a scientific talk, a TED talk, or a pitch and fully understanding and connecting with the entirety of the material. You feel alive, and you feel *smart*! Unfortunately, most presenters don't take the time to make sure their audience comes away feeling smart. Instead, their audience often struggles to interpret their complex jargon and follow along, as the presenter seems to have a goal of sounding incredibly smart.

The second the audience is confused or the message is unclear, they'll pull away from the talk, in part because they no longer feel connected and smart. If they start trying to connect the dots that you have not clearly connected, you are at risk of losing them, as the logic train is beginning to go off the cliff side.

To keep your audience engaged, you want to create an easy-to-understand presentation flow where the conclusion of each slide intuitively brings the audience to a question addressed on the next slide.

For example, if in my presentation, I tell you that we had some amazing impact with our product or platform, the customer or the VC is going to want to know how this was achieved. There is some creativity and testing involved in how I'll accomplish this, such as A/B testing, but my goal is to make sure my messaging, graphics, and story on this first slide get the person to think about the next slide and message. This will help you truly get your message across, and in some cases, it will be helpful in closing a sale or round of funding.

Early drafts of the logic flow can simply be putting the title of each slide or topic on a notecard and laying them all out in front of you. Now start to mess around with the arrangement and organize them in a way that flows well. Test the flow on friends and family. After you have the higher-level topics and slide titles in order, we can start to build in other visuals and materials that support your logical progression. Start to write out the other bullet point or two that you may make on your slide, and check for the logical flow again. With a test subject, show them a slide you have created, and ask them what they would expect to come next. If you have run this test a few times on your slides, and you see that your audience guesses right, then it is safe to say you have passed the test. The slides should not only flow logically, but should also naturally craft an exciting and engaging story (discussed in a separate chapter).

FIGURE 10.1

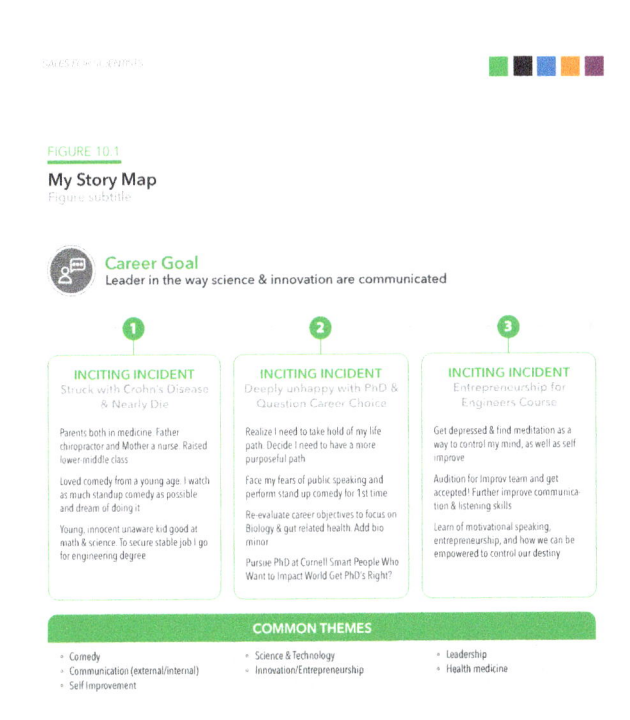

My Story Map
Figure subtitle

Career Goal
Leader in the way science & innovation are communicated

① INCITING INCIDENT
Struck with Crohn's Disease & Nearly Die

Parents both in medicine. Father chiropractor and Mother a nurse. Raised lower-middle class

Loved comedy from a young age. I watch as much standup comedy as possible and dream of doing it

Young, innocent unaware kid good at math & science. To secure stable job I go for engineering degree

② INCITING INCIDENT
Deeply unhappy with PhD & Question Career Choice

Realize I need to take hold of my life path. Decide I need to have a more purposeful path

Face my fears of public speaking and perform stand up comedy for 1st time

Re-evaluate career objectives to focus on Biology & gut related health. Add bio minor

Pursue PhD at Cornell Smart People Who Want to Impact World Get PhD's Right?

③ INCITING INCIDENT
Entrepreneurship for Engineers Course

Get depressed & find meditation as a way to control my mind, as well as self improve

Audition for Improv team and get accepted! Further improve communication & listening skills

Learn of motivational speaking, entrepreneurship, and how we can be empowered to control our destiny

COMMON THEMES

- Comedy
- Communication (external/internal)
- Self Improvement
- Science & Technology
- Innovation/Entrepreneurship
- Leadership
- Health medicine

Figure 10. Crafting Logic Flow for Presentation

Another neat trick to make your audience feel smart and special is by creating callbacks within your talk. A callback is when you tie early aspects of your presentation to the middle or end of it. This is helpful because your audience will recognize what you're talking about, and you'll pull them back in if you've lost them. When they see that callback connection, *they will feel smart*, which is what you want. You don't want them to feel confused and unengaged.

Preparation and Knowing Your Audience:

Practice, practice, practice. While it would be nice to walk into a big talk or presentation and not have to spend any time preparing, the more memorable speeches in history (like Steve Jobs' iPod talk) were expertly crafted and practiced to the point where they became second nature. There is no one-size-fits-all for presentations, but practice and hard work are always crucial elements. Additionally, make sure to practice your talk out loud. Everything is easier to remember and sounds better in your head. But when your words enter the real world and you are in front of an audience, the meaning and intonation and flow of your words change.

While it is important to practice your talk, do not be tempted to memorize it. Rather, **"memorialize"** it. This means no reading from the slides, but instead remember the 1 to 2 key points you have to make and how you want to say them. If you make the mistake of trying to memorize and are simply reciting every word, that means you have thousands of opportunities to get stuck, miss a word, or mess up the memorized flow; there is no room for improvisation or resilience in case you accidentally skip a slide or invert your key points.

A few other key tips in preparing for your talk:

- **Time your presentation ahead of time**. Be aware if you get lost while presenting or if you fly through it and speak fast. You are given a certain amount of time, so you want to use this time as effectively as possible. If you hope for the best and wing it, then it's almost a guarantee that you will either run out of time and fail to get to the key point or have too much time after, leaving you and your audience awkwardly waiting out the clock.

- **Listing titles and a critical path**. The first step for creating a great presentation is creating a list of all slide titles. Make sure each title addresses the one message you would like to make. Remember, it can become easy to add more noise and words to slides because all your points seem important and fun. But don't worry about slides until you have your titles and the key components of a story and have checked for the linear flow of your message. Run the 10-title progression past a few friends or colleagues; it should take only about 60 seconds. Define your critical path. If you had only titles and only one message to create, does your story logically get you there?

- **Simple and Visual Slides**: Use clear visual imagery and descriptions while using words to engage other senses to help them remember. Use a font size that stands out, most likely greater than 32. Using a large font size will also force you to limit how many words go onto your slides. The best slides have one simple image or visual and maybe a few words of text—that's it. When we try to include too much, it becomes distracting, and your message can become convoluted. Keep it simple. Realistically, you can give your talk without visual aids, and it will still be compelling and logical.

If you're giving a scientific talk and a graph is part of the slide, use bold and contrasting colors and use large text for axis titles. Label the title of the plot to be informative, and make the entire graph legible for an audience member in the back of the auditorium to see.

- **One Message Per Slide**: Put the one sentence you want to communicate as the title of each slide. Communicate one concept only per slide. If something else is important enough to communicate, then create an extra slide. Use a few words and visuals to complement the title and make it easier for them to grasp that one message. This applies to scientific talks as well. Sometimes there may be slide limits but the more you pack into a slide, the less your audience will grasp and remember.

- **A / B testing.** While this is extremely important and common in other parts of business like marketing, this may not be something you are too familiar with from the scientific or engineering perspective. A/B testing is essentially just forming a hypothesis and running a test to see the impact of one type of messaging versus another. In this scenario, we will be A/B testing our presentation material, slide titles, and slides themselves to see which versions are more easily understood so we can communicate the right messages. When using this in crafting a presentation it is not necessary to get hundreds of data points, but a few will go a long way.

- **Know your audience**. Instead of thinking about yourself and what you need, pay close attention to your target audience. Stop saying what you feel you need to say and start saying what they *need* to hear. To do this, you need to know your audience and what life looks and feels like for them. Ask yourself questions like: Why are they there? What challenges are they facing today? What are they passionate about? Where do they live? What is their education level? What do they hope for? What would excite or entice them? Until we can see the world through their lens, we will struggle to fully engage them in our talk. Remember, fish can't see water.

- **Choose the Right Stories.** The stories you weave into your talk will be the glue that holds your talk (and you) together. Stories serve multiple purposes in a presentation; they:

  1) Engage your audience

  2) Connect you to the present moment

  3) Add life to your facts

  4) Naturally bring energy to your words.

  Choosing the right stories is like selecting the right spices for a special dish. We can have all the appropriate images and visuals and words, but without a warm, loving story to connect it all, it feels like a pile of ingredients clumsily clumped together. Make sure to tell stories you can personally connect to and are passionate about. If you are not passionate about your story or the presentation, then why should your audience be? And if you feel you are stuck with a dry topic you don't care about, then find a story or a perspective on the topic that can bring life and passion to it. Craft your story with all the essential concepts we outlined in the book so you leave your audience feeling satisfied.

- **One Purpose:** Every presentation is delivered for a reason. Do you know why you are giving your talk? You should, and it needs to incorporate the one overarching message you make during your presentation. While you practice and even when you deliver your talk, make sure to keep a visible reminder of this message and what's at stake on hand. Every word, image, and story should be in service to your purpose and overarching message. When possible you should also compare and contrast the world if your purpose is or is not accomplished.

- **Call to Action**: Every great presentation must require something of their audience, even if just to take on a new way of thinking. The call to action "CTA" doesn't have to be a physical action or require much effort. In many situations, the CTA is a simple request of your audience to reflect or to challenge a preconceived idea or notion of the world.

Define your CTA while preparing your talk so it is woven in and is in service to your overarching message.

Starting Your Talk to Engage

We only have one first impression, and it is just as important in presenting as it is in dating. A nervous stutter, awkward phrasing, or fumbling with the slide quicker can quickly extinguish the audience's confidence in you. And once you lose them, you'll be fighting an uphill battle to get them back. And who could blame them? They've been sequestered to a chair and forced to listen to your talk against their will at work or at a conference. I've seen it estimated that at any time, your audience member can have over twenty different important topics scrambling through their brain, meaning that you are already low in priority.

To engage your audience from the start, there are a few essentials. The recipe I always use involves a simple acronym: B-W-W or Bang, Why, Who.

Bang, for starting with a memorable phrase that seems off-topic. **Why**, for making sure you answer your purpose or what has inspired you in this field or work. And **Who**, for what makes you the right person to be delivering this message. If each of these are included in your talk, you will be off to a great start.

The first words that come out of your mouth during a pitch, presentation, or sale should stick with the listener and dictate the first impression. Why waste them on norms like, "Welcome to the event that you already know you are at" or "Hi my name is X, and I would like to tell you about Y"? What a boring and uninspiring way to lose your audience. Let's face it—whether you are a scientist, an engineer, or a salesman, we are all fighting for people's limited attention and time.

A first impression is your chance to create a memory. Start with a Bang! Something profound, unique, ear-catching, and memorable that will cause your audience to stop and look up.

I deployed the Bang, Why, Who tactic when I won the 3-minute thesis competition at Cornell University. I knew I would be battling against hundreds of other science-related pitches and had to find a way to stand out. While I could have talked about how there is a need for research to solve Osteoarthritis and 1 in every 2 people get it at some point in their life, it was going to be one fact among a sea of other health and important facts. Instead, I tried to think of a statement that was relevant to my work, but only I would actually know why it is related. I started my talk with, "I've spent hundreds of hours collecting poop from mice." Weird right? It was also true (it was the best way to analyze the gut microbiome for research). By starting like this, I inspired curiosity from the start, and the audience remained engaged to find out how this was related to their expectations.

Here's a good recipe for a BANG statement: make sure it's unexpected and only you understand its relevance (humor also doesn't hurt!). Test it out on a few people and gauge their reactions.

The second essential component is a WHY. What is your WHY for this endeavor? Why are you here? Why did you start doing this research or create your company? There are numerous reasons WHY; just make sure you have one. A strong WHY allows your audience to see the world through your perspective. And if your audience can see the world through our lens, you are far more likely to be memorable and leave an impression on them, as they view every slide through this lens.

The third essential component is WHO. You are qualified, talented, and the right person to deliver your talk—and the audience needs to know this. I typically like to include this detail after the BANG and interspersed with my WHY. Catch their attention with the BANG, then tell them WHO you are and WHY you are doing this. You should not spend much time on the WHO part but think about your story and 1-2 defining moments in your life. You want to make sure these facts are highlighted and connected throughout your talk. If you need a little help with this, ask yourself questions like what is so exciting about you or why fate has brought you, the right person, to the place you will be presenting.

Here are a few of my favorite ways to start a talk:

- I love including a great first line or joke that I know I will get positive feedback or engagement from the audience. Check the chapter on humor for a few tips on integrating a situational joke or humor. I've found that humor from the start tips momentum in my favor.

- Starting with a question. The question can be about the scale of the problem or a topic indirectly related to your presentation. One way to do this: address your audience with a question at the beginning that you will (eventually) help solve with your talk. For example, if you give me 10 minutes of your time, by the end you will be able to give me a five-word description of your life's purpose. This is a way to grab their attention and give them an award for their attention.

- Share a fact that shocks the audience. Most days we drone on through an endless sea of meaningless information. Research statistics or facts relevant to your topic. Maybe you can describe the sheer amount of time and effort involved in your discovery. Perhaps you can offer an analogy of how many golf balls can be fit into your company's new manufacturing facility. Something that will *awaken* the audience and grab their attention from the start.

- Start with a personal story about when this topic or theme became important to you. So for you as a researcher or scientist, what is the first time you cared about science?. If it works well, then you are helping the audience have this same experience right now.

Excitement During Your Talk

Now that you have crafted your talk, there is one last step... *Delivering* it. The greatest talk not delivered is the same as the greatest meal prepared by the greatest chef, just to be thrown into the trash. Think of your presentation as a performance piece or a play. Plays are rehearsed thousands of times before opening night. So let's follow a few simple steps to make sure the delivery goes smoothly:

- **Breath**: Focus on your breath before, during, and after. The breath will be your center during the entire presentation. Just like in meditation, your breath will slow you down in a good way and force you to focus on the present moment. If you can keep track of your breath, then you can remain relaxed.

- **Vocal Variety**: We want to avoid monotone, boring, dry, vocal expressions that put audiences to sleep. That is why we need to vary our vocal expressions throughout our talk. I recommend picking a few words ahead of time to inflect your voice on. Choose words and phrases that inspire emotion and feeling inside your body.

- **Putting on a Show and Movement**: Create some theatricality and have fun with your presentation. Use physicality and act out what you are doing. Maybe it is taking a drink like the character in your story, or maybe, it is lassoing the competition. Use gestures to punctuate and signal change or momentum in your story. Your **movement** across the stage will add to the show if you can tie your movements to meaning in your talk. Maybe you can move across the stage from right to left to indicate the passage of time. Whatever you do, be careful, as constant undirected movement without a purpose will be distracting.

- **Read the Room and Improvise**: Despite all the preparation in the world, your talk may not hit the crowd perfectly, and you may need to audible (improvise) like an NFL quarterback reading a defense. This requires you to scan the room and the faces in the crowd to see if you are engaging them. If you find that the majority are not making eye contact or nodding their heads, or are beginning to snore, do not continue droning on until your time runs out. Stop and reconnect. Say something bold. Call people out directly. Refresh and make a scene about the gravity of the situation. Do not waste your moment and your preparation; re-engage. This may require a bit of improv, which simply means saying YES to the moment while adding something new to the next moment. You may want to take an improv class or two to practice what this looks and feels like.

- **Get the Audience Talking in the Beginning**: In a more intimate presentation (like in a sales situation), plan to engage directly with the audience within the first few minutes. You want your customer to be the one talking for 90% of your first sales calls. Best presentations in sales never end up going through all the slides.

Visualizing Your Successful Presentation

Visualizing can be an incredibly powerful tool to instill good outcomes in your mind and to ensure a successful presentation. Our minds can allow us to plan and see our entire talk before it occurs. On planet Earth, humans are a unique species because we have the ability to think abstractly, and project into the past as well as into the future. Think about the last time you were feeling stressed, worried, or frantic about an upcoming presentation. Think about all the sensations you felt in your stomach and brain and all over your body due to the worry and stress over everything you thought and visualized could go wrong. Those feelings you felt in advance of the presentation coursed through, and you wasted countless hours of brain power imagining this event... but in reality, you never actually delivered the presentation yet! These feelings in advance likely influenced the quality of your presentation, on some level. But the brain is just as powerful and able to influence the quality of something like a presentation if we rehearse greatness and success in our heads rather than worry and stress.

While this power of thinking in the the future can hurt us, we will use this instead to our advantage. We can create success through *visualization*. This is your secret weapon and what has allowed me to give amazing, award-winning presentations every time I've needed to. Visualizing may sound hippie or spiritual, but in reality you are already doing it before presentations, just not in a way that helps you. It happens all the time with our minds, every single day, and even while we are sleeping and dreaming. Before any big presentation I give, I close my eyes the night before and spend however long I need to visualize success, going through every detail, phrase, and gesture. I visualize that it is incredibly well-received by the audience, they are smiling, I am smiling, and at the end, they all clap. Go through and visualize all the emotions and thoughts you may have before, during, and after. What questions will you be asked by the audience? What might you be thinking about while on stage? You are essentially simulating this presentation *for* your brain. Our brain is so powerful in simulating the experience that we are actually "giving" the presentation that the quality of our visualization will directly affect the future outcome of what we want to experience. You may be skeptical, but at least try this technique once (while doing your best to put away your skepticism during your visualization). I can guarantee it will result in a more successful and enjoyable presentation.

Here's another secret that will give you confidence during your talk: Your audience wants you to succeed...maybe even more than you do. No one wants to watch you crash and burn on stage. Instead, they want to be inspired and motivated. So, the next time you are feeling nervous, know that everyone in front of you wants you to succeed.

SELF-REFLECTION EXERCISES

Slam Dunk Presentation Time

For this exercise, review one of your previous or upcoming talks. Infuse this presentation with questions, interesting facts, and a personal story. This is a time-intensive process and may take a few days. Save both the original and the new slides and accompanying script. Give the talk twice, once in each format. How was the audience's response? Were they following you the whole time? How did you feel delivering the new version? Did you feel more confident? Was it just generally easier since you practiced and crafted the flow? Take note. Great presentations are not only much better for the audience but for the presenter as well.

Visualizing Your Talk for Success

Instead of looking at a worst-case scenario, we will look at what an amazing, award-winning presentation looks and feels like. Start by writing down some of the emotions, fears, and negative thoughts you may feel before a talk. Next, close your eyes and visualize what it would be like to give a well-received presentation and to have confidence the entire time.

Are you smiling in it? Is the audience clapping? How is your posture? How do you feel on the inside? Are you calm?

After you have spent time visualizing the beginning, middle, and ending of your talk, write down everything you thought, felt, saw, smelled, and any other sensations during that visualization. Post-visualization, you should feel a little (or a lot) energized.

Finally, look at the fears and negative thoughts you wrote down initially. Do these feel as powerful or relevant now that you have spent time thinking about the positives and an amazing talk?

Bang, Why, Who

Think of the last talk you gave or an upcoming talk. How were you planning on starting off? Was it something basic like, "Hi, my name is, and I do blank?" Let's change it up and start with the BWW method! Come up with a **Bang** for the talk, a **Why**, and a **Who** that can be worked into the first few sentences. Use the tips above to make sure the *Bang* is seemingly off-topic to attract and engage the curiosity of the audience. Ensure your *Why* really allows the audience to grasp concepts through your perspective while experiencing your emotion on the topic. Let them know *Who* you are and how you are the best person to speak on the topic. Test it out on your next talk or with a few friends.

EMOTIONAL INTELLIGENCE AND YOUR AUDIENCE'S PERSPECTIVE

The Receiver of Your Presentations

When delivering a talk, consider the following questions to dive deep into your audience's mindset:

- What is the setting of the presentation? Is it at a large conference, an intimate board room, or an office?
- What does your audience hope to gain from the presentation?
- What is the educational level of your audience? What training and experiences might they have?

- Will your audience have any concerns about your topic that you need to address up front?
- Are there any cultural considerations or topics that could be misunderstood given diverse backgrounds?
- What does your audience need to know about the topic to fully digest your message?
- What will your audience do with the information?
- What emotional state is your audience, and how do you want them to feel after the presentation?
- Who else is in the room? Are they peers, managers, leaders, or other?

COMMON MISTAKES

- **Death by detail**: The thought that your work and ideas are revolutionary, and everyone will want to know every detail. Maybe you enjoy it, but it will be excruciating to your audience if you don't refine what you say and how you say it. Avoid losing your audience by only including minimal details needed to explain the concept. If the audience wants to know more, they can ask!
- **Not understanding your audience**: The whole point of our presentation is for our audience to learn or be influenced, so we better make sure we understand them. Often, we fail to dig deep in our understanding of our audience and do not frame our presentation around their perspective.
- **Using jargon to sound smart**: Many of us love to use the biggest and fanciest words and acronyms that all the "smartest" people in our industry know. It's not your job to sound smart; it's your job to make the audience feel smart. If you're the only smarty-pants in the room, then your audience won't care what you have to say. Ditch the jargon and acronyms to make sure everyone in the room understands what you're saying.
- **Losing them from the start**: Often, we dive deep into the "meat" of the presentation, assuming our audience will be excited and fully understand the context. This rarely

happens. Plan a hook from the beginning, or you will lose your audience immediately.

- **Taking the audience for granted**: You assume that just because you're in front and presenting that the audience should be happily listening and following along. Unfortunately, I know far too often that the audiences have a million other things on their mind and can lose focus or interest at any second. Too many presenters will just push through even though they see the audience members' eyes rolling back in their heads. Don't take it for granted that the audience will listen. Plan how you will engage the audience, and if you see nobody listening or interested, stop and get them back on track.

PUTTING IT ALL TOGETHER

Presentations are not simply a series of slides thrown together but an opportunity to craft a story that can compel your audience to make meaningful changes in their behavior. A presentation can change lives, save businesses, be the difference between success and failure, end or start someone's career, teach vital information, and do a million other important things. Spend the time and effort to appropriately prepare and deliver your presentation, because if you don't, you may as well leave the slides blank because what you're offering will have no impact on your audience. If you can master the art of presenting, you will notice life-changing results in your career and life.

CHAPTER 7

STORYTELLING

STORYTIME

"A story so compelling it can last generations"

Stories have been a part of humankind for thousands of years. Cave drawings are examples of early human stories. Stories are how we remember recipes or crucial facts, why we love songs, and how legends are born. The ability to narrate a story that engages your listener, invokes an emotion, and causes some internal reaction followed by a real-world action is a superpower. Simply your words can change someone's perception or action around something for the rest of their life. If that is not a superpower then I don't know what is!

A story can be so impactful because it can be easily shared and spread across a population like a virus. This is even more true now in the age of social media and the internet, where anyone can share their message globally in less than a second.

In the years I spent pitching venture capitalists to raise money for my company, it did not matter how amazing our product was or how big of a problem we were solving; if it wasn't memorable or simple enough to be shared verbally, then no one bought into what we were selling. Venture capitalists are meeting dozens of companies per week, and at the beginning of each week they would review what exciting companies they met with. As a startup, even if you had a great pitch, you are trusting that somebody else can explain it as well as you did. If you have a crisp story and narrative, it can be shared easily and you know that the message will be preserved when a venture partner tells the rest of the team later that week.

The same idea around storytelling applies to other areas of sales, business, and research. If you're at a conference presenting your research or presenting at a lab meeting, it is imperative to have a clear, memorable, and shareable story. Think about the goal of your research. One of the main criteria we see for grants and scholarships from the National Science Foundation, is to **disseminate our findings**. You can say whatever you want about our research, but at the end of the day you want others to share and have an opinion that is as close to yours as possible. If all they have are some scattered facts and numbers and no narrative, well then they will not be able to weave it together in their head, nor be able to weave it for them to share your message.

When a leading venture fund asked for a founder to share their fundraising tips on their podcast, they came to us because they could still tell us about that pitch even two years later. We told them a unique story that was fresh, with a unique industry, comical stories, and a meaningful impact to remember. We were in a new industry they had not been familiar with. Being completely NEW was memorable, and stayed fresh in their minds. If it sounds similar or there is no cohesive story, but rather a series of facts and figures, it will just fade away in their brain. This is the power of story, being able to meet only once, and have an important person still remember your story after listening to thousands of others.

WHY YOU SHOULD CARE & WHAT IT UNLOCKS

Stories are the most powerful way to influence people and the world. Every time we interact with another human, we are telling a story in some way. Maybe it is about our day, or what happened yesterday, or our hopes and dreams. If you can master the art of storytelling, your personal and professional lives will evolve, growing leaps and bounds. Your confidence will also grow because more people will be drawn to you and value your opinion and time.

PERSPECTIVES:

A Scientist's Perspective

As scientists storytelling is a subject we likely have never associated with our personal or professional life in any capacity. The last time we thought about storytelling was when we were reading Harry Potter or when our grandparents were reading to us. Scientists will feel there's really no need or place for storytelling in science, or in our professional life. Stories are for kids. Science and facts are what are important, and these alone are what people care about. As a result, scientists spend little to no time trying to craft a story in our daily lives, or in our presentations. This is a big mistake and misconception, and instead should be one of the first things scientists are taught before learning all the technical details they love.

A Salesperson's Perspective

As salespeople, we know that customers don't buy products; they buy stories. They invest in a vision of a better future. Sure, data points help paint the picture, but it's the narrative around the data that truly resonates, that compels action. Scientists often focus on presenting clear, unbiased data - a noble pursuit in the world of research. But in business, especially where disruptive technology is concerned, people yearn to believe in a story of change. Instead of being impressed by a technology that's 74.3% more effective, they're moved by the story of how this technology disrupts the status quo, how it revolutionizes their operations, and how it propels their company ahead in the market. This doesn't mean we disregard data. Far from it. We just frame it within a compelling narrative, transforming cold facts into a dynamic story of triumph. We show customers not only how our product works, but more importantly, how it can rewrite their story of success.

STATS THAT MATTER

- A study by the National Storytelling Network found that people are 22 times more likely to remember a story than a statistic. This demonstrates the power of storytelling to engage and persuade potential customers.
- According to a survey by Forbes, 84% of consumers say they are more likely to buy from a company with a story they can relate to.
- Another survey by the American Marketing Association found that 63% of consumers say they are more likely to buy from a company they feel has a strong brand story.

QUICK WINS

Embrace Storytelling and the Structure

There are 5 key components to a successful story.

1. Character and the current state
2. Inciting incident and goal
3. Turning point and challenge
4. Conflict and climax
5. Falling action and resolution

Make sure to connect and engage strongly in the beginning. Just like in film or TV; if you don't care about the character, you don't care about the story or message.

YOU HAVE A STORY AND BRAND

One of the most powerful first things you can do for your career is to reflect on important events from your life—where you have been and where you want to go—and use this to build a brand or origin story for yourself. Once you have crafted your story, think about a few key words that help summarize your brand so that people remember you after they've interacted with you. Your brand story, once perfected, can help you influence the world around you and help others understand why you are working in your field, why you started this company, or why they should care about you and your work.

DIGGING DEEPER

Scientists Can Learn From Movies

When crafting your stories, take some guidance from the movie world. Take structure from dramatic story writing for film, TV, or theater, such as:

1. Beginning with normalcy
2. Using an inciting incident
3. Creating a turning point (must change path)
4. Adding conflict
5. Offering a resolution

Told another way: There are really five key components of a compelling story:

1. Character
2. Goal
3. Challenge
4. Conflict
5. Resolution

Let's break down the five key components of a story that screenwriters use when crafting a memorable movie narrative.

Character and defining the current state

Introduced at the very beginning of every great story is a *likable* protagonist or main character. The initial information about the main character should state who they are and why the audience should care about them. The information should be presented so the audience builds a bond with this character and better understands their wants, interests, needs, and challenges. It should be written in such a way that the audience feels for the main character as they go through the unfolding journey. We care for iconic characters like Forrest

Gump, Luke Skywalker, and Bruce Wayne because we are presented with information that helps put their journey into perspective. The movie *Up* offers an outstanding protagonist introduction, giving insight into that character's current state and challenges simply through visuals. If you haven't watched it, I suggest you watch the first few minutes to see how much you learn about a character and how much you care about them.

Make writing a likable main character your goal when you craft your stories. It will cause the audience to form a bond with that character and care more about the content to come after. This is exactly why we need to craft who we are as a character, so our audience (an business partner, colleague, investor, or customer) cares about the journey and the story we are telling.

During the early part of your story, you also need to establish the current state of the world. This means defining how the current process is performed, the current understanding of scientific theory, or any other relevant information that allows you to set up a starting point for your story so you have a final destination to get to by the end of your story. The audience should be able to follow along as your main character works to create a new or better world by the end of the story. For a scientist or an engineer, this can be the story of how you first became interested in and inspired by this scientific topic as a kid. For an entrepreneur, it can be a struggle you went through in the past that humanizes you, or it could involve all the failures you had to endure before this idea.

Make sure you spend a good amount of time creating a likable main character and setting up the current state of the world; it is the backbone of a successful story.

Inciting incident and Goal

In a film, TV show episode, or book, the inciting incident is major action or conflict that causes a change of path for the main character. The inciting incident is often a dramatic moment to draw the viewer into the film and the journey ahead. This moment puts the protagonist into the main action of a story. For *Saving Private Ryan*, the death of the other three brothers is the inciting incident that puts the rest of the story in motion to save the last brother. In *Monsters Inc*, it is the adorable little girl confronting Sully and challenging his reality and opinion of his world. Your main character will need a clear, high-stakes goal that connects to the inciting incident. The story you tell will follow the main character's pursuit of this high-stakes goal. For scientists and engineers telling the story around their research, this incident can be a failed experiment that led to a different path. For an entrepreneur, your inciting incident can be a painful experience that hit you in the face and made you aware of a larger problem in the world that has not yet been solved. A good inciting incident helps your audience emotionally connect and care about the journey to come.

Turning point and challenge

It is now the point in the story where the hero has had some early success, only to be confronted by a roadblock or challenge. The main character must rise to the challenge and find a way to overcome the odds against him. Sometimes there can be more than one turning point that subjects our hero to a new environment or new conditions, but eventually, there is a turning point that forces him to fully commit to a path and never look back. Reflect on any movie you have seen, and you will see that no hero simply chooses a path and effortlessly glides to the end.

For a scientist or engineer, this can be the portion of the story where you thought you had figured everything out, but in reality, the problem was much bigger, or your solution did not work yet. For an entrepreneur, this can be where we thought we had figured out a solution to the problem, but it actually required a more complex technology that you have now built.

Conflict and climax

The climax is the most exciting part of the story. The main character battles the ultimate villain. In movies and fiction books, every detail beforehand leads to the climax. Imagine Batman fighting the Joker or Luke battling Darth Vader. This portion of our story needs to reflect the epic circumstances and the high stakes for the hero.

For a scientist, this can be the moment where, against all odds, you make the discovery or breakthrough in your research that will change the world and your career forever. For the entrepreneur, this is where you finally deploy your idea in the real world, and it actually solves the problem you were hoping it would. When done right, the audience will be able to empathize with and root for the main character because the writer has taken them on the same long, arduous journey.

Falling Action and Resolution

The main action of the story has now ended, and the story threads begin to wind down. Eventually, all the loose ends of the story are neatly tied back together, leaving the audience with a satiated feeling. Think of how every Disney movie somehow magically resolves all the issues, and you feel as if things are at peace for the main character. For a scientist, this is where your breakthrough has been verified and now has started curing the once-fatal disease in large numbers. The scientist looks back at his lab with a smile and turns off the light as the lab fades away into the credits.

The only difference between us and the movies is we usually tell our stories to persuade the audience to take action. So, we need to **add a call to action** at the end of our story.

Congratulations! You now know how to tell a story just like the writers of the million-dollar movies you watch on the big screen.

YOUR STORY

Whether or not you believe there is an aligned narrative driving your life, you want your audience to believe that when you present to them and tell your story. You applied to X job or built Y company because of the perfect alignment of the universe and your journey to get here.

As Steve Jobs said, "You can only connect the dots looking back, not forward."

As you look to craft your story, think about some key questions: What major successes or life events come to mind first? What would your family and friends say if you asked them the same question about yourself? What is your unifying narrative? Is it a trait, a passion, a goal, a skill, or a destiny that everything seems to revolve around?

Your story will also change based on your audience and your situation. The majority will stay the same, but you'll need to frame it a little differently depending on the audience's perspective. Here are a few examples of how my story evolves and changes based on my audience:

- **Growing Up in the Industrial Warehouses**: I grew up working in industrial warehouse facilities until I got severely injured falling from a truck. This made me aware of what workers in these jobs are exposed to daily, and I saw the lack of appreciation toward them. This inspired me to learn how I could help these workers become safer and happier. So, I went to school to pursue a Ph.D. in biomechanics so I could understand how to keep workers safe.
- **Growing Up in Medicine**: My entire career has been focused on health and injury prevention. My father was a doctor, and my brother an orthopedic surgeon, exposing me to the power of biomechanics. I pursued my Ph.D. to understand biomechanics and apply it to prevent injuries in the blue-collar workforce.

Your story will continue to evolve and grow, so make sure to update your resume, CV, and story map and configure for to every job or presentation you are leading!

Why is it so important to have an unbelievable personal story? Everything starts with you—research, a company, a job, or your family. If you are going to work on sales and communicating for your business or your research, you need to be able to tell an engaging and compelling story about yourself. It is also a great way to get to know yourself!

Crafting your story will unlock future value for you in a variety of contexts. You don't want the first time you answer the question "Tell me about yourself?" to be in front of the key investor. Perfect your origin story so your words have a big impact on your audience.

MAPPING OUT YOUR LIFE STORY

The best starting point for building your story starts with mapping out your life story and key events that have happened along the way. All of us are on a boat in life, but we sometimes fail to check where the currents and winds are pushing us. We'll use my life as an example.

First, we'll start with a list of notable ideas, topics, and components of my life story and how I have been able to weave them together. Note that I have tried to keep each point brief. It is not necessary to write in complete sentences for this exercise.

- Raised in a lower-middle-class family in Long Island, New York.
- Growing up, I was an athlete who played several sports such as basketball and tennis.
- Always had an emphasis on science and math through high school.
- Lots of exposure to the medical field with my mother as a nurse and dad as a chiropractor.
- I was extra shy growing up. Could not do presentations. Could barely speak to strangers.
- I got Crohn's disease and nearly died in the hospital from a related infection.
- Did stand-up comedy for the first time to fight my fear of public speaking after nearly dying.
- I went back to school to pursue a biology minor and a Ph.D. at Cornell on the gut microbiome and orthopedics.
- I founded my first company to use wearable technology and AI to help prevent injuries in the workforce.
- After 6 years, my passions changed, and the need to support other life things like marriage, buying a house, and having a kid emerged.
- I switched roles to work on product management for health and hardware products at Fortune 25

After you have listed all your defining moments, traumas, decisions, and anything else that comes to mind, you can now start to connect the dots. A good way to do this is to write all the ideas onto Post-it notes and lay them on a table. Organize them chronologically but try to group them if you start to see different patterns emerge. For example, for me, clear themes that emerged were medicine, wearables, and health. I could see my parents' background and careers, my brother's career, and the main trauma I had to overcome was health/medical related. I could see that public speaking and shyness were some of my weaknesses, that I have chosen to focus on in my life, and I have been able to overcome those over the years. Check out my example life story map below:

Figure 10. Example Life Story Map

What are the obvious candidates from my map for each aspect of the story? It will depend on our audience, but let's assume I am applying for a technology leadership position at a health tech company.

First, we will look at each component of the story, with a long and short version provided:

1. **Character and defining the current state:**

    a. **Long**: My fascination with health and medicine started at a young age because my parents worked as a doctor and a nurse. My first memory of understanding how amazing the human body is was cutting my foot on glass as a toddler. Blood was gushing everywhere, and as a 3-year-old with little life perspective, I thought I was going to bleed out and die. But I got a Band-aid, and to my amazement, I healed. From this point, I became enamored with medicine and the power of the human body. I excelled at everything biology-related: classes, tests, and labs.

    b. **Short**: As a young kid I was enamored with health and medicine, with my earliest memory being amazed how I was able to heal myself after a cut..

2. **Inciting incident and goal:**

    a. **Long**: While in college, working my way through an engineering degree, I became extremely ill. Fevers and chilis and a hard lump in my gut that was painful to touch. The day I went to the ER was a day that would change my life forever. The doctor came out while I waited on the stretcher in the hospital hallway with my parents, delirious with a fever and without food. My mom rushed to him to hear the diagnosis. I was unable to hear him over all the loud noises of the hospital, but I could read his lips as he said, "I'm sorry, but he has Crohn's disease." I was crushed and heartbroken, especially after seeing how much pain and impact this disease had on other members of my family. I decided that day that I would need to devote my life and career to improving health, and that while medicine had come far, there was still much farther to go.

    b. **Short**: It was not until I was in college though that my ilfe and motivations changed forever. I nearly died from Crohn's disease after a sudden thirty day battle in the hospital and dropping nearly 40 pounds. After surviving I vowed to dedicate my life to advancing health and medicine.

3. **Turning point and challenge:**

    a. **Long**: I decided to pursue a Ph.D in biomedical engineering and focus on research topics directly related to my disease, and my passion of sports. While working on my Ph.D. I learned I wanted to have a bigger impact faster on the health of those around me. I realized I could leverage and build new technology to help millions of people injured in the workplace. So, for the next several years I began building a wearable platform to help preserve worker health. Initially, I built the company in the evenings and on weekends and went full-time once I graduated. We recruited a small team of outstanding talent. We secured some grants and initial seed funding. Things were going surprisingly smoothly for a little while. As we were scaling, fundraising was becoming an increasingly important focus because our initial funds had evaporated while building the technology and paying contractors. We were looking at some upcoming deadlines for bills and last opportunities for customers if we were unable to secure additional funding. A clear and present challenge had emerged for me.

    b. **Short**: I pursued a Ph.D in biomedical engineering, to advance science in the areas that mattered to me most and to gain essential skills. I soon founded a wearables and AI platform to preserve manufacturing workers health and have the large impact I wanted.

4. **Conflict and climax:**

    a. **Long**: Looking at the bank account and our monthly burn, along with a growing list of investors who had told us "no"; the death of the company and our dream was becoming more imminent. So, I put together a plan and a due date by the Christmas holiday. If we could not secure funding by then, it would mean the end of our livelihood and company. For the next two months, I spent every waking moment pinging investors, pitching, meeting, and going through due diligence to try and get funding. Unfortunately, we were not getting any firm commitments and needed a lead for this funding round to happen. We had less than a few thousand dollars in the account, our credit cards were maxed out, and we hadn't paid ourselves for months. It felt as if the end was near. But we kept pushing, calling, and trying to find funding. Just when we had nearly given up, I got a special call from a special investor who wanted to lead our round. The company and our mission was saved. We couldn't have asked for a better holiday present.

    b. **Short**: The livelihood of the company was a constant struggle for funding, resources, and time. We soon hit a seemingly impossible wall of No's and funding near its end. I spent every piece of energy in my body to close an investor. And just when the it was the darkest, we somehow got that special call to save the day.

5. **Falling Action and Resolution:**

   a. **Long**: With the money now in our account, we were able to keep the lights on and make a few key hires in the following months. Our dream lives on, and our technology continues to help many. We learned firsthand that it's always darkest before dawn.

   b. **Short**: With the funding now in our account we were able to continue to expand and keep our employees.

A condensed version of this story for a 60 second overview would be closer to:

"My life has always been centered around health, with the most impactful moment being when I nearly died from an unexpected battle with Crohn's disease at 20. This was a key turning point as I would decide to dedicate my life to the advancement of human health and medicine, starting with a PhD at Cornell in biomedical engineering. I excelled at interdisciplinary research, advancing the field of orthopedics and the gut microbiome with industry first publications in the field, but still longed for a bigger impact for human health sooner. As such, I founded my first company, leveraging AI and wearables to protect the health and safety of the millions of workers in the industrial sector. My shining moment when we were near out of funds during the pandemic, was securing funding at the very last moment before our bank accounts ran out"

Have multiple versions of your story with different lengths and for different audiences. I recommend a 15-second, 30-second, 60-second, and several-minute version, depending on the format of your presentation. Having multiple lengths of our story was useful when fundraising for startups, depending on the context of an investor meeting. Sometimes, all I had was an elevator ride, and other times, I had a ten-minute staged presentation.

CRAFTING YOUR BRAND

The best day to start working on your brand is yesterday. You might think only companies like Nike and Lululemon need to worry about branding. But you are already projecting a brand with how you speak, what you wear, what you show on social media, and many other things that influence how others perceive you. Your brand is a glimpse into your perspective on the world.

As a scientist or technical person, you may believe there is only one dimension to your brand—that you are technical and smart. However, there is much more, and we can use these extra dimensions to our advantage if we are aware of them. My brand has evolved greatly over the years, especially after I realized I was subconsciously advertising one for many years already. Words that I try to have others perceive me as are: intelligent, compassionate, innovative, open-minded, motivated, and hard-working. I'm not saying I am lucky enough to have others think of me in that light all the time, but at least it's my goal.

A few examples of brand archetypes I've seen over the years with scientists and engineers I've worked with.

- The *Yes Man* who pleases everyone but has no opinions of his own.
- *Mr. Practical* who follows rules and stays within boundaries at all times
- *Troublemaker* who picks at the issues and unsolved problems
- The *Innovative Genius* who always thinks outside the box
- *Unstoppable Fighter* who will never give up
- *Questioner* who needs to understand all the details and the "why" before doing anything
- *Calculated Thinker* who does not react to anything
- *Negative and Critical Engineer* who only sees the issues or problems with a new idea.

Your brand is not a stagnant line of code; it will continue to evolve and change as you go through life. Take the time to assess your brand every year and see that your actions and successes are building toward the brand you want because it can be easy to lose sight of the bigger picture.

When a company builds a brand from the ground up, they consider the following:

- **Navigation**: How the brand helps consumers choose them versus the other brands in the market? For you this means thinking about how your brand can stand out compared to your peers and competitors for the same job or funding.
- **Reassurance**: Does the brand help consumers to feel and understand the intrinsic value of their offering. For you this means thinking about how your brand communicates your biggest strengths.
- **Engagement**: Does the brand use distinctive imagery and language to help consumers identify with the brand. For you this means thinking about how your brand will connect with your target you are trying to reach for your sale, your idea, or your research.

It is also helpful to develop a mission statement for your brand you can refer back to at any point. For companies, a brand mission statement includes a brand's purpose, objectives how it will serve those objectives, and sometimes a comment on the future vision.

As you craft your brand and mission statement, make sure they follow many of the communication rules we have outlined in this book such as:

- Simple
- Memorable
- Authentic
- Addresses your Why

HUMOR

The way to anyone's heart and a great way to influence someone is through laughter. With humor on your side, the door is now open for you to convince and persuade. In most presentations and professional settings, everyone wants to cut the intense atmosphere with a knife. This means that jokes are appreciated, and the bar for a successful joke is lower than it would be in a comedy club. I've often found myself laughing at jokes that I thought were maybe 2 out of 10, but in an awkward professional setting, I laughed like it was Jerry Seinfeld at Madison Square Garden in New York City.

- If you can make someone smile and laugh, it will make you more likable and cause the listener to be more open to your message and fully receive it
- Think about the last few times someone made you laugh. Did or do you like this person? Try to find anyone in your life who has made you laugh and see if you don't like them. I'm almost certain you'll have some kind of affection for anyone who makes you laugh.

Humor tips and different types of jokes:

- Puns, though sometimes cringeworthy, are an easy opportunity because you're relying on simple wordplay. See if any keywords in your presentation could be said differently or be replaced. Is there a common phrase where one of your words actually fits into the memorable expression.
- Jokes can be created about the minutiae of research or the topic, as well as that you have chosen to spend your entire life on a small, specific topic. For example, an agricultural researcher on apple genetics could say something like, "I've chosen to spend my life focused on apples rather than curing cancer."
- Comparing one topic to another. "I've always found my research to be similar to making breakfast." Then make a few points on how they are similar. "It works better in an oven on 350 and often needs to be paired with a fat source."
- Something about the stage or the venue. E.g., if the sound stops working. "I asked them to make sure the microphone would be glitchy today. I wanted to seem more relatable. Is my plan working?"
- The topic. Often, something along the lines of, "I know what you are thinking, another talk on ___."

- Exaggerating the size or scale of something with hyperbole. "As large as a whale." "I raised enough mice for this experiment to colonize the Moon."
- About presentations. Think about the audience's perspective. Presentations (besides yours!) are boring, repetitive, and a chore. Playing with this commonly accepted reality is often a recipe for success.
- Self-deprecating humor can also be an easy one.
- Humor can be humanizing. Oftentimes bringing in a light comment about your family can prove a great way to get a laugh and demonstrate your relatability. Maybe it is about not getting enough sleep. Maybe it is about the wrong tie or shoes because your kids replaced them.

If, for some reason, you are unable to think of or create your own relevant joke, there is still hope. With a little bit of creativity, you can use pre-existing material and package it into your presentation using the techniques below:

- If you can't think of an actual joke, you can get a little help by doing some internet research. Don't plagiarize from your favorite standup comic. But there are plenty of dad jokes and old standup in the public domain.
- Finding a relevant and funny comic strip, image, meme, or video can lighten the mood and even help tell the story you've worked so hard to put together. Think of the moment of discovery, is this like a kid from a movie trying ice cream for the first time. Think about the chaos in your lab working all night until you found that discovery, was it like a car crash in a James Bond movie.

Note: never make fun of another person for their physical attributes, beliefs, relationships, or anything else that could seem off-color. Use your humor wisely to grow relationships, influence others, and communicate your messages.

EXAMPLE: BEFORE / AFTER

Situation: During the startup selling phase, we found it difficult to communicate our ideas and the value we could provide to a customer—especially when selling a complicated and technical product. We tried for nearly a year, focusing on the technical details and how the technology worked thinking this would grasp the customer and impress them. We were wrong. Among many challenges in our sales pitch, we did not tell an easily remembered or relatable story. Once we realized this gap and the power a story could add, we decided to switch gears.

- **Before**: Sales pitch focused on the technology. "We leverage wearables to collect millions of data points on your workers and the way they move to understand injury and safety risks". We would explain the sensors in the wearable, how the data is sent, and how the analytics and machine learning work. But still, after all this, we found that our customer was not interested nor did not fully understand. So, we added more figures and diagrams and went into further detail. We soon discovered that all these efforts were steps in the wrong direction.

- **After**: We knew something needed to be done to improve sales and connect with potential customers. We reviewed all our greatest successes and examples of our platform working with customers. We then decided if we used one of the success stories from a leading customer, that this could potentially work. So, we took a relatable example, made a character out of the frontline worker for our audience to care about, and created an easy story to remember and retell about how our technology detected one type of movement with the wrist. The team used this information to modify the job and the way the worker performed the tasks. We removed all the parts about how the technology worked and only talked about it if the customer asked specific questions. It worked like gold, and we were able to close more customers in the next six months than we had the whole last year.

SELF-REFLECTION EXERCISES

Map Out Your Story

Mapping out your life story can help you define your end goals. It allows you to connect the dots looking back while offering you a better understanding of your trajectory and how you may want to pivot or change course in the future. Do you see any common themes that tie the beginning of your story to the present moment or your future goals?

What Words Should Be Associated With You

What adjectives come to mind when you think of Steve Jobs? What about Elon Musk? Maybe words like "bold," "visionary," or "leader" come to mind. That is most likely no accident. People like Jobs and Musk control their narrative and take actions to build upon it further.

Now it is your turn to take control of your narrative. Make a list of ten adjectives you would like to be described as ten years from now. Are you going for a detailed-oriented analyzer who follows rules to the tee or an innovative creative genius? Once you have your list, try to narrow it down to only three or four because taking action on these will be easier and more sustainable. Next, list actions that you have taken in your career so far that support your listed adjectives. How many can you list? Finally, think about what actions you can take in your career and life to support being associated with those adjectives.

Funnying Up Your Past

Let's pull up your last scientific or sales talk. Did you include any jokes in your presentation? Even talks on serious topics or with important stakes can benefit from humor, but you might have to work a little harder at it.

Your goal for this exercise is to add two jokes to the presentation you have selected. Maybe you need to add a joke to the introduction to give yourself a little more confidence and to gain the audience's attention. Perhaps you want to add a pun or wordplay during the deep technical portion of your talk. Be creative and experiment! Make sure to practice several times in front of a live audience before presenting, or your perfectly crafted joke may fall on unhappy listeners.

COMMON MISTAKES

- **Missing story sections**: A great story has all of the essential components. Skipping the character section can be the most impactful mistake of all, as your audience will not care to engage with the rest of the storytelling.

- **Not personalizing the story to each audience**: Even the best of stories will still sometimes need to be modified to better align with a particular audience. Don't make the mistake of a one size fits all approach.
- **Unrelatable Character**: Sometimes even though a character has been introduced, they are still unrelatable, meaning your audience will not engage. Make sure to consider varying perspectives and the full context to make sure your character is relatable.
- **Poor Delivery**: An amazing story presented with low energy and no emotion makes the entire story worthless. A good story should be practiced and developed in parallel with the delivery.

PUTTING IT ALL TOGETHER

Storytelling is a necessary skill to practice and then to master. It's useful in nearly every interaction in your personal and professional life. It is foundational in being able to communicate your ideas and to influence others. A perfectly crafted story evokes emotions from the listener, catches them from the start, and is easy to follow the entire way through. The more you practice the concepts of storytelling, learn from others, and pay attention to these concepts in the movies and shows you watch, the more your knowledge base will grow. Storytime is no longer just for kids; it's for adults too!

START SELLING

It is that time we release you back into the wild. It can be easy to revert back to our old comfortable ways immediately. But we are trusting that you have done the exercises in this book, you have written out your goals and a plan to track your progress, and most importantly you have your WHY, that will keep you motivated. The tips in this book are not rocket science, but more us helping to shine and guide the light on well known facts learned in other industries. Scientists are equipped with everything they need to be a successful seller, but just need a little bit of encouragement and direction to get there. We thank you for taking the time to work on yourself, to stretch, and for taking the first big step to expanding your impact and success. Now get out there and start captivating the world with your innovations and persuading millions to believe in your ideas.

For the first part of many of our careers as scientists, we are focused on the nitty gritty details of everything, being judged on 100% perfection of our knowledge of facts and figures. We view that our success largely depends solely on accuracy and facts. For the next part of your career, where you focus on getting what you want and what you deserve and excelling in your career, the shift is away from pure facts and figures. Your success will now more largely depend on your relationships,, your network, people who remember your impression on them, your ability to present, to tell stories, to persuade and influence. If you fully step into your role as the selling scientist, the only way is up.

Made in the USA
Las Vegas, NV
20 July 2023